STARGAZING 2006

MONTH-BY-MONTH GUIDE TO THE NORTHERN NIGHT SKY

HEATHER COUPER & NIGEL HENBEST

HEATHER COUPER and NIGEL HENBEST are internationally recognized writers and broadcasters on astronomy, space and science. They have written more than 30 books and over 1000 articles and are the founders of an independent TV production company, specializing in factual and scientific programming.

Heather is a past President of both the British Astronomical Association and the Society for Popular Astronomy. She is a Fellow of the Royal Astronomical Society, a Fellow of the Institute of Physics and a Millennium Commissioner. Nigel has been Astronomy Consultant to New Scientist magazine, Editor of the Journal of the British Astronomical Association, and Media Consultant to the Royal Greenwich Observatory.

Published in Great Britain in 2005
by Philip's, a division of Octopus Publishing Group Ltd,
2–4 Heron Quays, London E14 4JP

Text: Heather Couper & Nigel Henbest (pp. 6–53)
Robin Scagell (pp. 61–64)
Philip's (pp. 1–5, 54–60)

ISBN-13 978-0-540-08789-1
ISBN-10 0-540-08789-0

Printed in China

Details of other Philip's titles and services can be found on our website at: **www.philips-maps.co.uk**

Title page: The Pleiades (Robin Scagell/Galaxy)

ACKNOWLEDGEMENTS

All star maps by Wil Tirion/Philip's, with extra annotation by Philip's.
Artworks © Philip's.

All photographs from Galaxy.
44 Yoji Hirose
32, 40 Damian Peach
1, 8, 12, 16, 29, 36, 48, 62, 63, 64 Robin Scagell
24, 53 Michael Stecker
61 Optical Vision
20 Dave Tyler

CONTENTS

The sight of diamond-bright stars sparkling against a sky of black velvet is one of life's most glorious experiences. No wonder stargazing is so popular. Learning your way around the night sky requires nothing more than patience, a reasonably clear sky and the 12 star charts included in this book.

Stargazing 2006 is a guide to the sky for every month of the year. Complete beginners will find it an essential night-time companion, while seasoned amateur astronomers will find the updates invaluable.

THE MONTHLY CHARTS

Each pair of monthly charts shows the views of the heavens looking north and south. They're useable throughout most of Europe – between 40 and 60 degrees north. Only the brightest stars are shown (otherwise we would have had to put 3000 stars on each chart, instead of about 200). This means that we plot stars down to 3rd magnitude, with a few 4th-magnitude stars to complete distinctive patterns. We also show the ecliptic, which is the apparent path of the Sun in the sky.

USING THE STAR CHARTS

To use the charts, begin by locating the north pole star – Polaris – by using the stars of the Plough (see April). When you are looking at Polaris you are facing north, with west on your left and east on your right. (West and east are reversed on star charts because they show the view looking up into the sky instead of down towards the ground.) The left-hand chart then shows the view you have to the north. Most of the stars you see will be circumpolar, which means that they're visible all year. The other stars rise in the east and set in the west.

Now turn and face the opposite direction, south. This is the view that changes most during the course of the year. Leo, with its prominent 'Sickle' formation, is high in the spring skies. Summer is dominated by the bright trio of Vega, Deneb and Altair. Autumn's familiar marker is the Square of Pegasus, while the winter sky is ruled over by the stars of Orion.

The charts show the sky as it appears in the late evening for each month; the exact times are noted in the caption with the chart. If you are observing in the early morning you will find that the view is different. As a rule of thumb, if you are observing two hours later than the time suggested in the caption, then the following month's map will more accurately represent the stars on view. So, if you wish to observe at midnight in the middle of February, two hours later than the time suggested in the caption, then the stars will appear as they are on March's chart. When using a chart for the 'wrong' month, however, bear in mind that the planets and Moon will not be shown in their correct positions.

THE MOON, PLANETS AND SPECIAL EVENTS

In addition to the stars visible each month, the charts show the positions of any planets on view in the late evening. Other planets may also be visible that month, but they won't be on the chart if they have already set, or if they do not rise until early morning. Their positions are described in the text, so that you can find them if you are observing at other times.

We've also plotted the path of the Moon. Its position is marked at three-day intervals. The dates when it reaches First Quarter, Full Moon, Last Quarter and New Moon are given in the text. If there is a meteor shower in the month we mark the position from which the meteors appear to emanate – the *radiant*. More information on observing the planets and other Solar System objects is given on pages 54–57.

Once you have identified the constellations and found the planets, you'll want to know more about what's on view. Each month, we explain one object, such as a particularly interesting star or galaxy, in detail. We have also chosen a spectacular image for each month and described how it was captured. All these pictures were taken by amateur astronomers. We list details and dates of special events, such as meteor showers or eclipses, and give observing tips. Finally, each month we pick a topic related to what's on view, ranging from solar eclipses to comets to space missions, and discuss it in more detail. Where possible, all relevant objects are highlighted on the maps.

FURTHER INFORMATION

The year's star charts form the heart of the book, providing material for many enjoyable observing sessions. For background information turn to pages 54–57, where diagrams help to explain, among other things, the movement of the planets and why we see eclipses.

Although there's plenty to see with the naked eye, many observers use binoculars or telescopes, and some choose to record their observations using cameras, CCDs or webcams. For a round-up of what's new in observing technology, go to pages 61–64, where equipment expert Robin Scagell shares his knowledge.

If you have already invested in binoculars or a telescope then you can explore the deep sky – nebulae (starbirth sites), star clusters and galaxies. On pages 58–60 we list recommended deep sky objects, constellation by constellation. Use the appropriate month's maps to see which constellations are on view, and then choose your targets. The table of 'limiting magnitude' (page 58) will help you to decide if a particular object is visible with your equipment.

Happy stargazing!

The start of the New Year is the most exciting time for appreciating the magnificence of the starry sky. Darkness falls early, and the nights are often frosty and clear. Centre-stage of the heavens is the glorious constellation of **Orion**, fighting his adversary **Taurus** the Bull – accompanied by trusty hounds **Canis Major** and **Canis Minor**.

These constellations include many of the sky's brightest stars – including the most brilliant of them all, **Sirius**, towards the south in Canis Major. Sirius forms one end of a great arc of stars lying to Orion's left, which stretches from **Procyon** (in Canis Minor) through **Pollux** and **Castor** (in Gemini) to **Capella** (in Auriga).

▼ The sky at 10 pm in the middle of January, with Moon-positions marked at intervals of three days. The positions of the stars are also correct for 11 pm

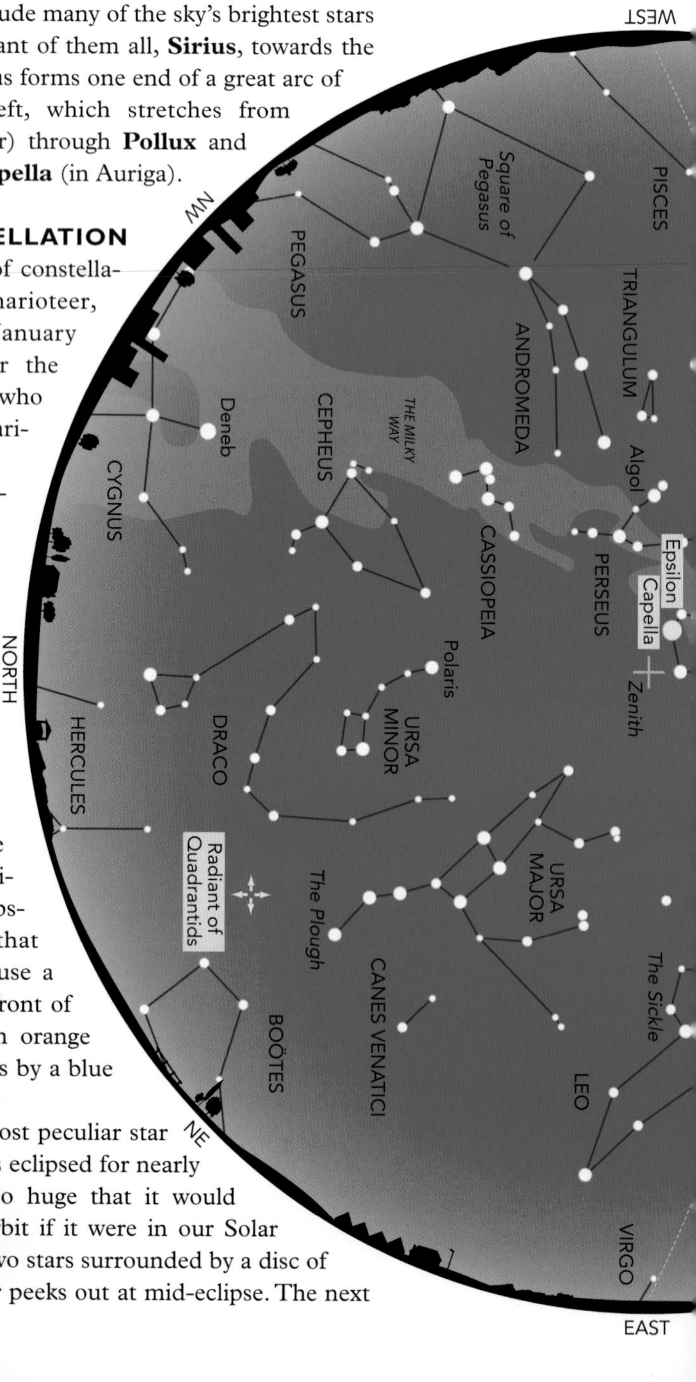

JANUARY'S CONSTELLATION

One of the most ancient of constellations, **Auriga**, the Charioteer, sparkles overhead on January nights. It is named after the Greek hero Erichthoneus, who invented the four-horse chariot to combat his lameness.

Capella, the sixth-brightest star in the sky, dominates Auriga – marking the Charioteer's shoulder. Its name means 'the little she goat', but this is a giant star over 150 times more luminous than our Sun. To its right is a tiny triangle of stars nicknamed 'the kids'. Two of the three are variable stars; they are not intrinsically unstable but are 'eclipsing binaries' – stars that change in brightness because a companion star passes in front of them. **Zeta Aurigae** is an orange star eclipsed every 972 days by a blue partner.

Epsilon Aurigae is a most peculiar star system. Every 27 years, it is eclipsed for nearly two years by something so huge that it would stretch beyond Saturn's orbit if it were in our Solar System. It may be one or two stars surrounded by a disc of dust, because the main star peeks out at mid-eclipse. The next eclipse is due in 2009–11.

at the beginning of January, and 9 pm at the end of the month. The planets move slightly relative to the stars during the course of the month.

When Auriga is overhead, binoculars or a small telescope will allow sight of three very pretty star clusters within the constellation – **M36, M37** and **M38.**

PLANETS ON VIEW

Saturn, on view all night long, shines brilliantly (magnitude –0.2) in the dim constellation of Cancer. On 27 January, Saturn is at opposition (in a line with the Earth and the Sun) and lies 1216 million km away. A small telescope will make out its rings, and possibly its largest moon. At around 7 pm on 25 January, Saturn moves in front of a star (HD 74050) that's slightly brighter than Titan – an exciting event for those with large telescopes.

Venus lies low in the evening twilight at the start of January. At magnitude –4.4, it outshines everything apart from the Sun and Moon. It ceases being the Evening Star on 14 January, when it swings between the Earth and the Sun to reappear in the dawn skies as the Morning Star.

Red Planet **Mars**, in the west between Aries and Cetus, fades rapidly this month, as the fast-moving Earth pulls away from it. Mars begins January at magnitude –0.6 but by the month's end has dimmed to +0.3.

Jupiter rises at around 3 am, low down in the southeast in Libra. Brighter than any star, it shines at magnitude –1.9. The four largest of its over 60 satellites can be seen with binoculars or a small telescope. On 13 January, Jupiter passes near the double star Zuben Elgenubi – a fine sight in binoculars.

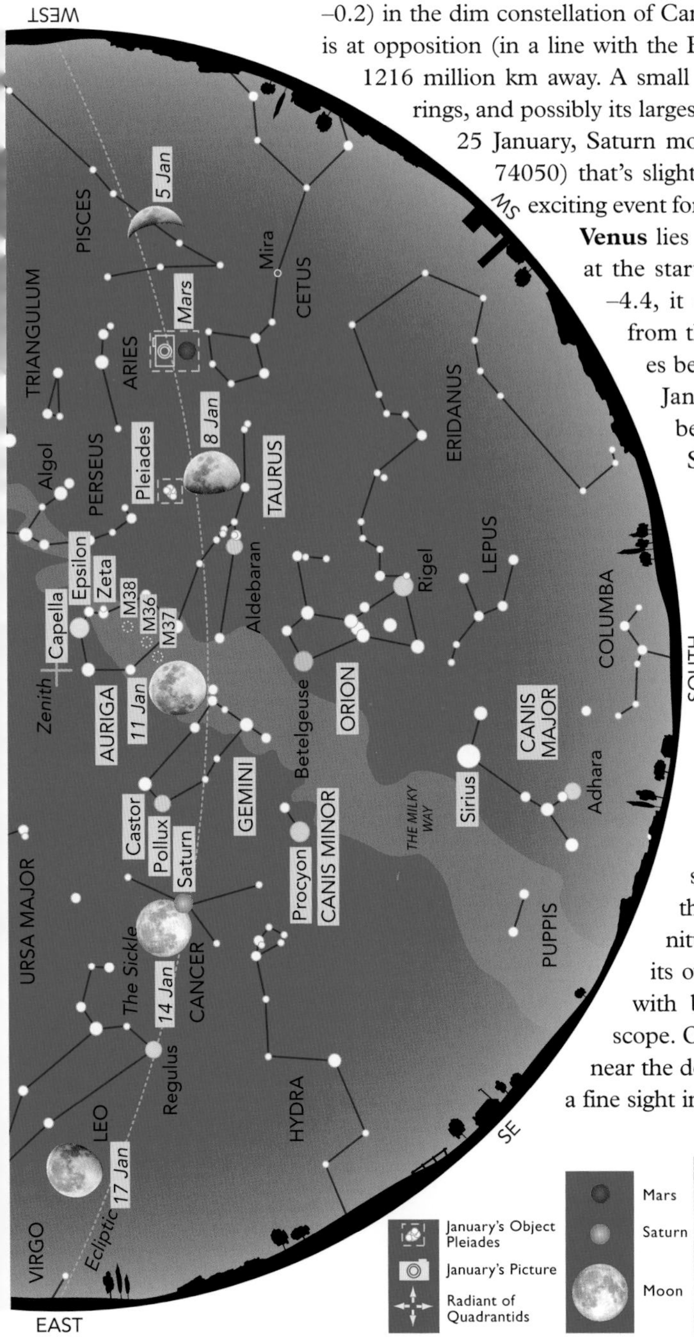

MOON		
Date	Time	Phase
6	6.56 pm	First Quarter
14	9.48 am	Full Moon
22	3.14 pm	Last Quarter
29	2.15 pm	New Moon

Uranus (magnitude +5.9) lies in Aquarius and sets at around 8 pm. **Mercury** and **Neptune** are too close to the Sun to be seen this month.

MOON

The narrow crescent Moon lies near brilliant Venus on the first two evenings of the New Year. On the night of 8 January the Moon (in its gibbous phase, between First Quarter and Full) is very close to Mars. It grazes the edge of the Pleiades star cluster the following night. On the evenings of 14 and 15 January the Moon lies near Saturn, and passes Jupiter on the mornings of 23 and 24 January. The thin crescent Moon is visible in the twilight with Venus on 26 January – this time in the morning sky.

SPECIAL EVENTS

3 January: It is the maximum of the **Quadrantid** meteor shower. These shooting stars are tiny particles of dust that burn up as they enter the Earth's atmosphere. Perspective makes them appear to emanate from one spot in the sky, the *radiant* (marked on the starchart). This is a good year for observing the Quadrantids because moonlight will not interfere: you may see up to 100 meteors per hour. Astronomers once believed that this stream of debris was shed by Comet Machholz, but the culprit is in fact an asteroid called 2003 EH1 (probably the core of a burnt-out comet).

4 January: Earth is at perihelion – the point in its orbit when it is closest to the Sun. The Earth travels around the Sun in an elliptical path, and on this date it lies 147 million km from our local star.

▲ *A composite of images of the Moon and Mars. It was captured by Robin Scagell on a 215 mm Newtonian telescope, using Kodachrome 64 film with an exposure of 1/60th second.*

JANUARY'S OBJECT

'A swarm of fireflies tangled in a silver braid' – this evocative description of the lovely **Pleiades** (or Seven Sisters) star cluster was coined by Alfred, Lord Tennyson in his 1842 poem

Viewing tip
It may sound obvious, but if you want to stargaze at this most glorious time of year, dress up warmly! Lots of layers are better than a heavy coat (for they trap air next to your skin), heavy-soled boots stop the frost creeping up your legs, and a woolly hat will indeed stop one-third of your body heat escaping through the top of your head. And – alas – no hipflask of whisky: alcohol constricts the veins, making you feel even colder.

'Locksley Hall', and how accurate it is. Very keen-sighted observers can pick out up to 11 stars, but most have to content themselves with six or seven. These are just the most luminous in a group of 400 stars, lying about 400 light years away (the precise distance is still debated). The cluster's brightest stars are hot and blue, and all of them are young – less than 80 million years old. They were all born together, and have yet to go their separate ways. The fledgling stars have blundered into a cloud of gas and dust; on CCD images this looks like gossamer, but the Pleiades are still a beautiful sight to the naked eye or through binoculars.

JANUARY'S PICTURE

On 8 January, **Mars** will be only half a degree away from the Moon in the sky – just a moonwidth. It will make a lovely sight, with the golden gibbous Moon so close to the Red Planet (as in this image). You will be able to watch the Moon moving relative to Mars, and the event will be easy to capture using a telephoto lens.

JANUARY'S TOPIC
New Horizons

On 11 January, an Atlas launcher will loft the New Horizons spacecraft on a journey to Pluto – the only planet that has not yet been visited by a probe from Earth. New Horizons' itinerary will take it close to giant Jupiter, whose mighty gravity will speed it on its way to encounter Pluto and its moon Charon on 14 July 2015. Instruments on board will map the smallest planet (at 2274 km in diameter, even smaller than our Moon), observe its thin atmosphere and look for additional moons and rings in the system. Afterwards – if New Horizons has enough power from its plutonium-powered radiothermal generators – it will examine some of the other smaller Pluto-like bodies in the Kuiper Belt of outer icy asteroids. And it may finally answer the question: is Pluto a genuine, stand-alone planet, or just a big Kuiper Belt Object?

This month, **Saturn** is still dominant, riding high in the sky in the constellation of Cancer. But there are signs that spring is on the way. The brilliant winter stars are beginning to drift towards the west as Earth travels around the Sun on its annual journey.

Capella, the brightest star in Auriga, has moved from the overhead position that it claimed for much of winter, and **Orion** is shuffling towards the wings. But February is an excellent time to see **Canis Major**, dominated by **Sirius**, the sky's brightest star. From northern latitudes Sirius is very low in the sky, although this can make for sensational visual effects: you'll often see the star twinkling and flashing all the colours of the rainbow. This has nothing to do with Sirius, but a lot to do with Earth's atmosphere. Looking at the stars through the atmosphere is like looking at your surroundings from the bottom of a swimming pool. As the water shifts and ripples, the view is distorted. Earth's atmosphere similarly is always on the move, making the stars appear to twinkle. It also refracts starlight, creating the multicolour light show.

▼ The sky at 10 pm in the middle of February, with Moon-positions marked at intervals of three days. The positions of the stars are also correct for 11 pm

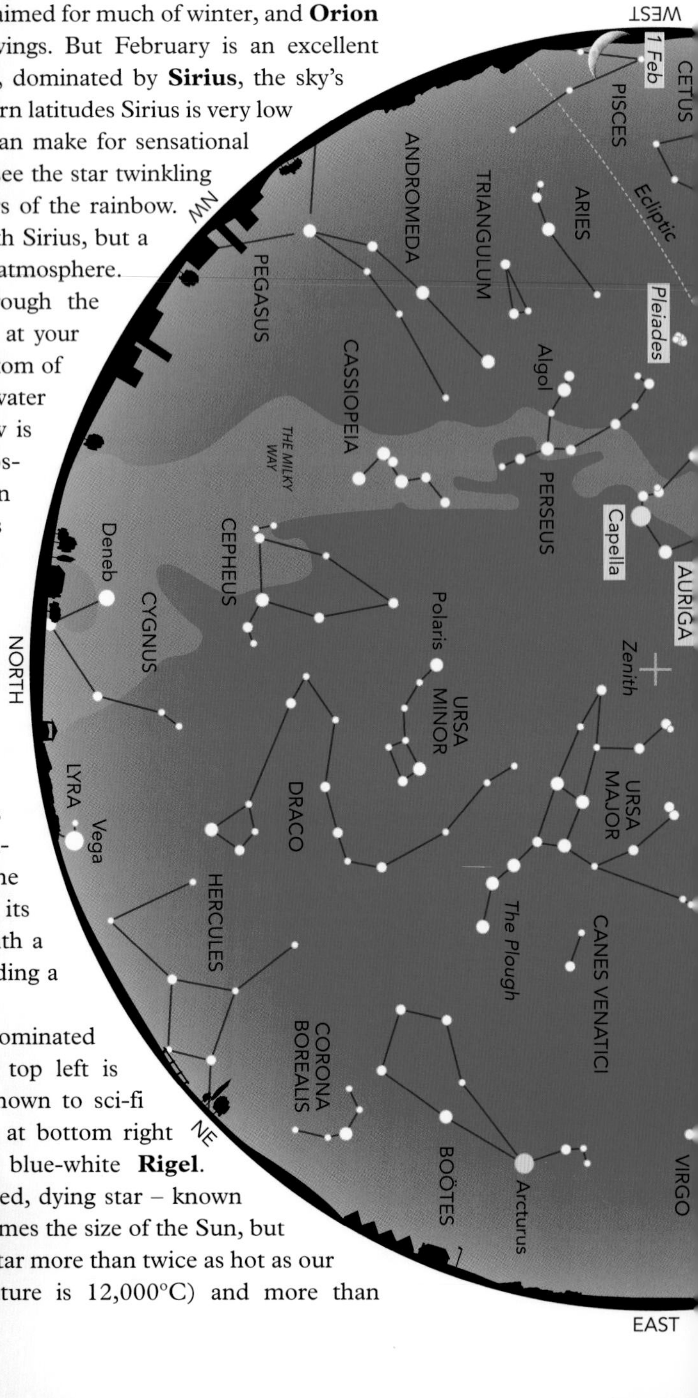

FEBRUARY'S CONSTELLATION

Orion – the ultimate hunter in mythology – is the most recognizable constellation in the sky, and one that actually looks like its namesake: a giant man with a sword below his belt, wielding a club above his head.

The constellation is dominated by two brilliant stars: at top left is blood-red **Betelgeuse** (known to sci-fi fans as 'Beetlejuice') and at bottom right the even more brilliant blue-white **Rigel**. Betelgeuse is a cool, bloated, dying star – known as a red giant – over 300 times the size of the Sun, but Rigel is a vigorous young star more than twice as hot as our Sun (its surface temperature is 12,000°C) and more than 50,000 times as bright.

t the beginning of February, and pm at the end of the month. he planets move slightly relative the stars during the course of e month.

The famous 'belt of Orion' comprises the stars **Alnitak** (left), **Alnilam** and **Mintaka** (right). Below it hangs Orion's sword, containing the glowing gas cloud of the **Orion Nebula** – a nursery of young stars.

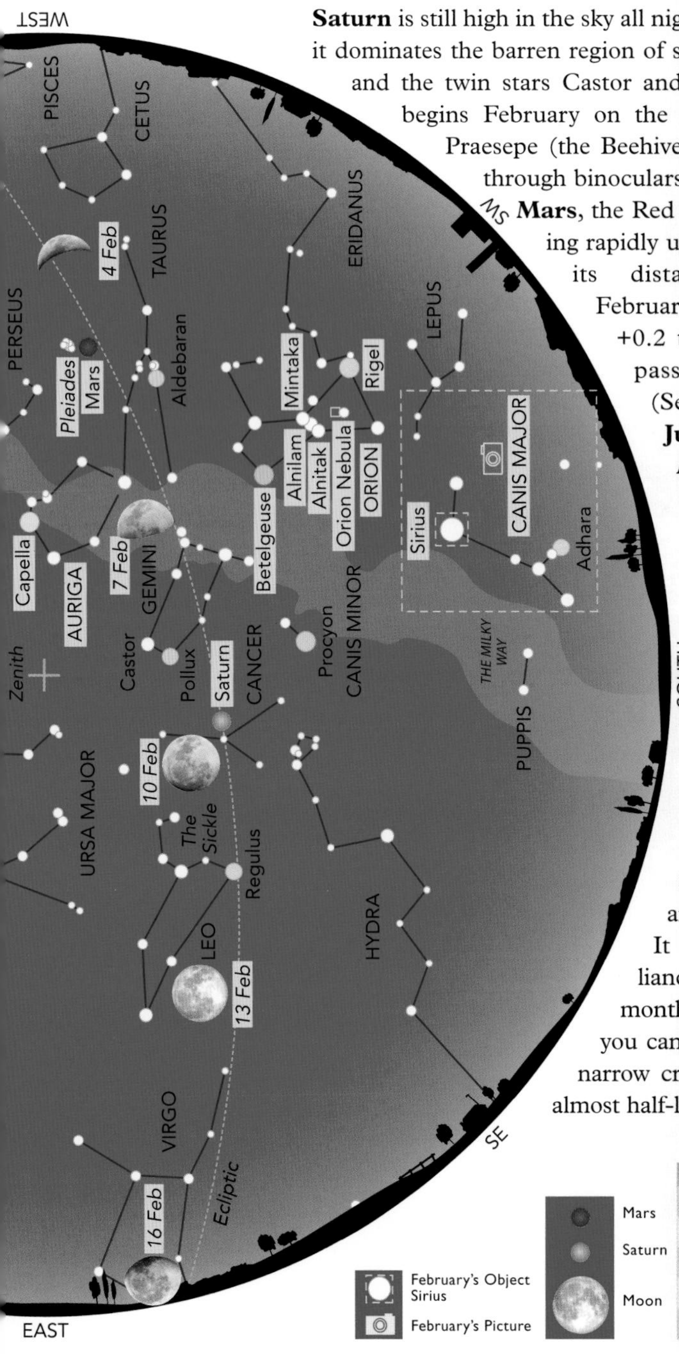

PLANETS ON VIEW

Saturn is still high in the sky all night long. At magnitude –0.1, it dominates the barren region of sky between Regulus in Leo and the twin stars Castor and Pollux in Gemini. Saturn begins February on the fringes of the star cluster Praesepe (the Beehive), making a gorgeous sight through binoculars or a small telescope.

Mars, the Red Planet, to the west, is moving rapidly upwards in the sky, fading as its distance increases. During February its magnitude drops from +0.2 to +0.8. Mid-month, Mars passes below the **Pleiades** (Seven Sisters) star cluster.

Jupiter rises at around 1 am. At magnitude –2.2, it's by far the brightest object in Libra. During the last week of February, if you are searching for its four biggest moons with binoculars or a small telescope, there will appear to be five! The interloper is the star Nu Librae, which lies very near Jupiter from 23 February to 16 March.

The glorious Morning Star – **Venus** – is rising at around 5 am in the southeast. It reaches its maximum brilliance, of magnitude –4.6, mid-month. Through a small telescope, you can see Venus changing from a narrow crescent in early February to almost half-lit by the month's end.

February's Object Sirius
February's Picture

Mars
Saturn
Moon

MOON		
Date	Time	Phase
5	6.29 am	First Quarter
13	4.44 am	Full Moon
21	7.17 am	Last Quarter
28	0.31 am	New Moon

This month offers the year's best chance to see the elusive **Mercury** in the evening sky: 24 February marks its greatest elongation from the Sun. From 20 February to the month's end, look due west as the first stars are coming out (at around 6.30 pm). Shining at around magnitude 0.0, Mercury is pretty bright but is easily lost against the glow of the dusk sky.

Uranus and **Neptune** are too close to the Sun to be visible this month.

MOON

The First Quarter Moon lies near Mars on 5 February, and Saturn on 11 February. On the morning of 18 February, the Moon passes very close to the 1st-magnitude star Spica in Virgo – from locations south of the UK, the Moon will pass right in front of Spica and occult it. The Last Quarter Moon is near Jupiter on the morning of 20 February. On the mornings of 23 to 25 February, the thin crescent Moon ducks down below brilliant Venus.

Viewing tip

When you first go out to observe, you may be disappointed at how few stars you can see in the sky. But wait for around 20 minutes, and you'll be amazed at how your night vision improves. The reason for this is that the pupil of your eye gets larger to make the best of the darkness. Observers call this technique 'dark adaption', and it also involves the increased production in the retina of a chemical called rhodopsin, which dramatically increases the eye's sensitivity.

FEBRUARY'S OBJECT

This is the month of the brightest star in the sky – **Sirius**. It isn't a particularly luminous star: it just happens to lie nearby, at a distance of 8.6 light years. The 'Dog Star' is accompanied by a little companion, affectionately called 'The Pup'. This tiny star was discovered in 1862 by Alvan Clark while he was testing a telescope; it had however been predicted nearly 20 years earlier by Friedrich Bessel, who had observed that something was 'tugging' on Sirius. The Pup is a white dwarf: the dying nuclear reactor of an ancient star, which has puffed off its atmosphere. White dwarfs are the size of a planet but have the mass of a star; and because they are so collapsed, they have considerable gravitational powers for their diminutive size – hence Sirius's 'wobble'. The Pup is visible through medium-powered telescopes.

FEBRUARY'S PICTURE

The constellation of **Canis Major** is here photographed from Tillingham, Essex. Brilliant **Sirius** is at the top of the picture. The lights in the foreground are from the North Kent coastline, some 60 km away. More needs to be done about light pollution so that our dark skies can be reclaimed.

FEBRUARY'S TOPIC
Mercury

Setting nearly two hours after the Sun, Mercury makes its best evening appearance this month. As Mercury is the innermost planet, never straying far from the Sun, getting a glimpse of the planet is quite rare – but now's your chance! (Rumour has it that Copernicus, the Solar System's 'architect', never saw Mercury because of mists from the Danube.)

Mercury looks like a bright, untwinkling star, low on the western horizon. But don't expect any revelations, even through a good telescope. The planet is only a little larger than our Moon – and it bears an uncanny resemblance to it. Mercury is covered in craters (although you need a space probe to see them), and crisscrossed with wrinkle-ridges like the skin of a dried-out apple – the result of the planet contracting after its hot birth.

NASA's Messenger spacecraft is on its way to Mercury, and is due to arrive on 18 March 2011. In 2011–12, Europe plans to launch a more ambitious probe, BepiColombo, which will map the planet thoroughly and land on its surface.

◀ This early-morning shot of Canis Major was taken by Robin Scagell on 14 December 1996. He used ISO 1600 film with an exposure time of about a minute.

The nights become shorter than the days as we head into spring. The official start of the new season is 20 March – the Vernal (Spring) Equinox – when the Sun moves over the equator on its way from the southern hemisphere of the sky to the northern hemisphere.

At the Equinox, day and night are equal, with exactly 12 hours between sunrise and sunset. Also on that day, the Sun rises due east and sets due west – wherever you may be in the world.

Because the Earth is tilted at 23.5 degrees with respect to its orbital path, the north pole points away from the Sun between September and March – hence our northern autumn and winter – and towards it between March and September, when we enjoy spring and summer.

The change of season is reflected in the sky, with the arrival of the spring constellations such as **Leo** and **Virgo**. And on 26 March, British Summer Time (BST) begins, when the clocks go forward by one hour.

Mars and **Saturn** still adorn the evening sky, while brilliant Jupiter and Venus are morning objects. March also promises two very different eclipses, of the Sun and the Moon respectively.

▼ The sky at 10 pm in the middle of March, with Moon-positions marked at intervals of three days. The positions of the stars are also correct for 11 pm

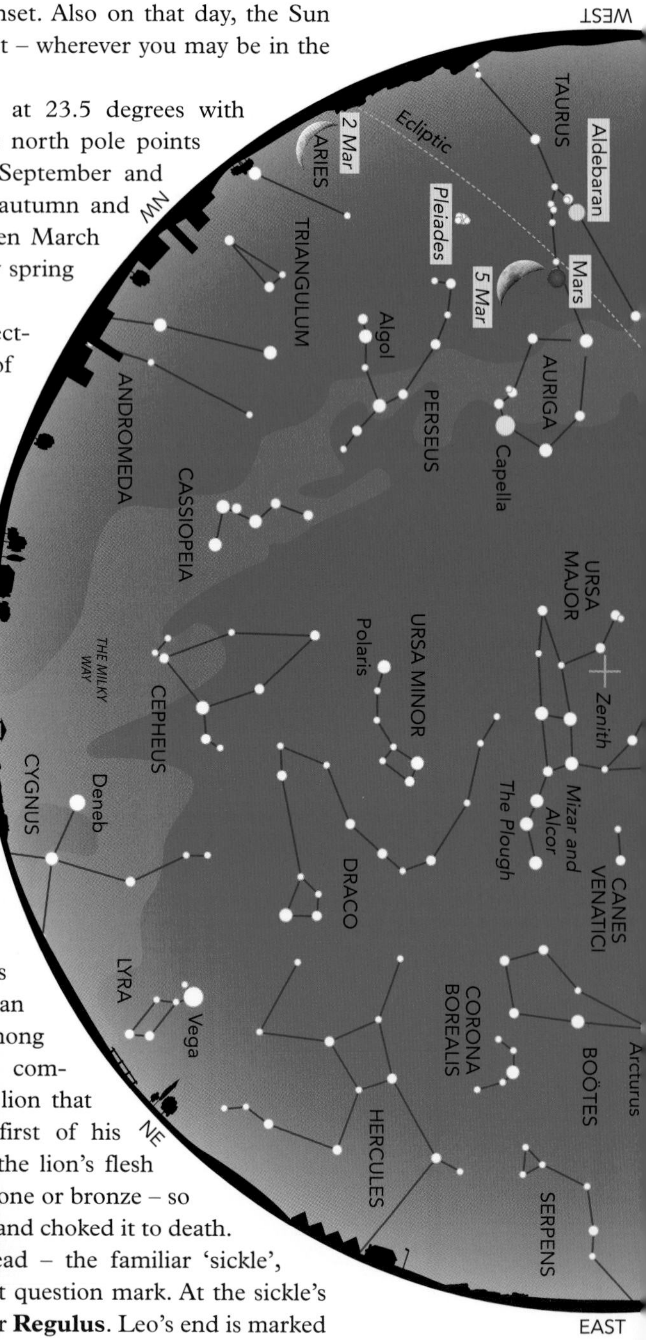

MARCH'S CONSTELLATION

Like Orion, **Leo** is one of the rare constellations that looks like its namesake – in this case, an enormous crouching lion. Among the oldest constellations, Leo commemorates the giant Nemean lion that Hercules slaughtered as the first of his labours. According to legend, the lion's flesh could not be pierced by iron, stone or bronze – so Hercules wrestled with the lion and choked it to death.

Leo is dominated by his head – the familiar 'sickle', which looks like a back-to-front question mark. At the sickle's base is the bright blue-white star **Regulus**. Leo's end is marked

t the beginning of March, and
0 pm at the end of the month
ıfter BST begins). The planets
ıove slightly relative to the stars
ıring the course of the month.

by **Denebola**, which in Arabic means 'the lion's tail'. Just underneath Leo's main 'body' are several spiral galaxies – nearby cities of stars like our own Milky Way. They can't be seen with the naked eye, but a sweep along the lion's tummy with a small telescope will reveal them.

PLANETS ON VIEW

At magnitude –4.5, **Venus** is a brilliant Morning Star, rising in the southeast at around 4.30 am (5.30 am BST). Through a small telescope, the planet appears to shrink in size as it pulls away from Earth, while its shape changes from a crescent to half-lit – like a miniature Last Quarter Moon. On 25 March Venus is at its maximum distance from the Sun (greatest western elongation).

Mars continues fading during the month, from magnitude +0.8 to +1.2. It starts the month near the **Pleiades** star cluster, forming a 'twin' with red giant star **Aldebaran** in Taurus (the bull). Mars moves rapidly upwards in the evening sky, ending the month between the bull's 'horns'.

Saturn is still prominent in the evening skies. At magnitude 0.0, it lies in Cancer and is visible for most of the night.

Jupiter (magnitude –2.3) lies in the faint constellation of Libra, and rises in the southeast at around 11 pm. For the first two weeks of March, Jupiter is still very close to the star Nu Librae (magnitude +5.2), and as a result seems to have five, rather than four, bright moons when viewed through binoculars or a small telescope.

MOON		
Date	**Time**	**Phase**
6	8.16 pm	First Quarter
14	11.35 pm	Full Moon
22	7.10 pm	Last Quarter
29	11.15 am	New Moon

March's Object
Regulus

Mercury, along with outer planets **Uranus** and **Neptune**, is too close to the Sun to be seen this month.

MOON

The First Quarter Moon lies near Mars on the night of 6/7 March, and it passes Saturn on the evening of 10 March. During the night of 18/19 March, the waning Moon is found near to brilliant Jupiter. On the mornings of 25 and 26 March, the crescent Moon shares the dawn twilight with the Morning Star, Venus. The Moon is also involved in two eclipses in March.

SPECIAL EVENTS

14/15 March, 9.22 pm–2.14 am: The Full Moon suffers a penumbral eclipse, as it moves into the outer regions of the Earth's shadow. Don't hold your breath for this eclipse: the whole of the Moon's surface remains partially illuminated by the Sun, and all that you'll see is a slight dimming of Full Moon. Many people won't even notice anything unusual.

20 March, 6.25 pm: The Vernal Equinox marks the beginning of Spring, as the Sun moves up to shine over the northern hemisphere.

26 March, 1.00 am: British Summer Time starts – don't forget to put your clocks forward (the mnemonic is 'Spring forward, Fall back').

29 March, 10.45–11.37 am: There is a total eclipse of the Sun (see March's Topic).

Viewing tip

This is the time of year to tie down your compass points – the directions of north, south, east and west – as seen from your observing site. North is easy: just find Polaris, the Pole Star. But the useful extra in March is the Spring Equinox, when the Sun hovers over the equator. This means that it rises due east, and sets due west. And at noon, the Sun is always due south. So remember those positions relative to a tree or house around your horizon.

◀ *In this beautiful picture of a partial eclipse, very similar to this month's event, the Sun is only five degrees above the horizon – hence the vivid red colours. Robin Scagell used a 400 mm telephoto lens, and Kodachrome 64 film.*

MARCH'S OBJECT

Regulus – the 'heart' of Leo the Lion – appears to be a bright but fairly anonymous star. Some 77 light years away, it's young (a few hundred million years old), 3.5 times heavier than the Sun, and it throws out 350 times as much energy as our local star. But recent discoveries have revealed it to be a maverick. It spins in less than a day – meaning that it has a rotational velocity of over one million km/h. If it were to spin only 10% faster, it would tear itself apart. This bizarre behaviour means that its equator bulges like a tangerine (its equatorial girth is one-third larger than its north–south diameter). To compound it all, its rotation axis is tilted at an angle of 86 degrees – which means that Regulus zaps through the Milky Way virtually on its side.

MARCH'S PICTURE

The Sun with a chunk missing! This is the kind of view of our local star that we expect to see from the UK on 29 March, when the Moon moves in front of the Sun and causes a partial eclipse. The 'missing bit' is where the dark disc of the Moon overlaps the Sun. *You must use eclipse goggles to view this event.*

MARCH'S TOPIC
Solar Eclipse

North Africa is the place to be this month. It's where you can catch a glorious total eclipse of the Sun, which will last for up to 4 minutes and 7 seconds if you journey to the border of Libya and Chad. Solar eclipses are caused when the Moon passes in front of the Sun and blocks off its light. By complete coincidence, the Moon and Sun appear to be the same size in our skies – the Sun is 400 times bigger than the Moon, but it's also 400 times further away.

Seeing a total eclipse is an incredible experience. The blazing Sun is replaced by a celestial dragon mask – a black mouth surrounded by a glowing halo (the Sun's outer atmosphere, or *corona*). The eclipse is total over a wide swathe of the Earth, from Brazil (where it starts) through North Africa to central Asia and Mongolia (where it ends).

Either side of the line of totality, the eclipse will be partial: the Moon still overlaps the Sun, but not exactly – and the beautiful corona won't be seen. From London, 25% of the Sun will be obscured. It's highly dangerous to look at the Sun even if it's a quarter covered-up, so drag out those eclipse goggles (checking they're undamaged!), and you'll see the Sun with a chunk taken out of it.

Two ferocious beasts ride high in April's sky: **Ursa Major** (the great bear) almost overhead, with **Leo** (the lion) below. Between Leo and the southern horizon sprawls **Hydra**, the water snake, which straggles over as much as 100 degrees – almost one-third of the way round the entire sky. A distinctive quadrilateral of stars marks out **Corvus** – a crow perched on the snake's back.

To the east (left) of Leo lies the ancient constellation of **Virgo**, represented as a Y-shape of stars extending from 1st-magnitude **Spica**. Above bright blue-white Spica lies the orange giant star **Arcturus** – a glorious contrast in colours when viewed through binoculars.

Mars and **Saturn** also grace the evening sky, while the brilliant Morning Star is Venus. This planet will be in the headlines in April, because the European spacecraft Venus Express will arrive – all being well – to probe the secrets hidden beneath Venus's all-enveloping clouds.

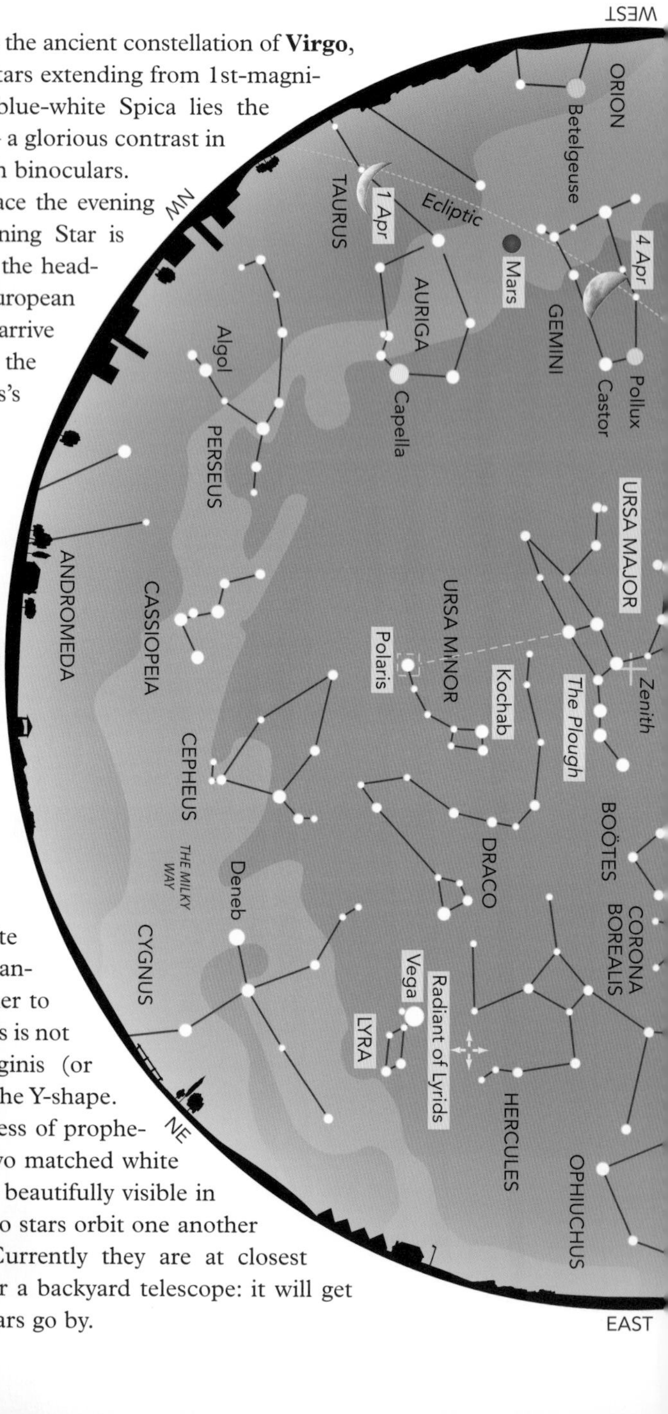

▼ The sky at 11 pm in the middle of April, with Moon-positions marked at intervals of three days. The positions of the stars are also correct for midnight

APRIL'S CONSTELLATION

The Y-shaped constellation of **Virgo** is the second-largest in the sky. It takes imagination to see the group of stars as a virtuous maiden holding an ear of corn (Spica), but this very old constellation has associations with the times of harvest (the Sun passes through the stars of Virgo in early autumn).

Spica is a hot, blue-white star with a close stellar companion. They're too close together to be seen in a telescope, but this is not the case with Gamma Virginis (or **Porrima**) – the 'centre' of the Y-shape. Named after a Roman goddess of prophecy, Porrima is a system of two matched white stars, which are – as a rule – beautifully visible in a small telescope. But the two stars orbit one another in a very eccentric path. Currently they are at closest approach and a challenge for a backyard telescope: it will get easier to split them as the years go by.

t the beginning of April, and 0 pm at the end of the month. he planets move slightly relative the stars during the course of he month.

The glory of Virgo is the Y-shape's 'bowl'. Scan it with a small telescope and you'll find it packed with faint, fuzzy blobs – just a few of the 3000 galaxies that make up the gigantic Virgo Cluster. This cluster is 10 million light years across, and its residents are also large-scale. Some harbour supermassive black holes in their cores, and spew out jets of gas at close to the speed of light.

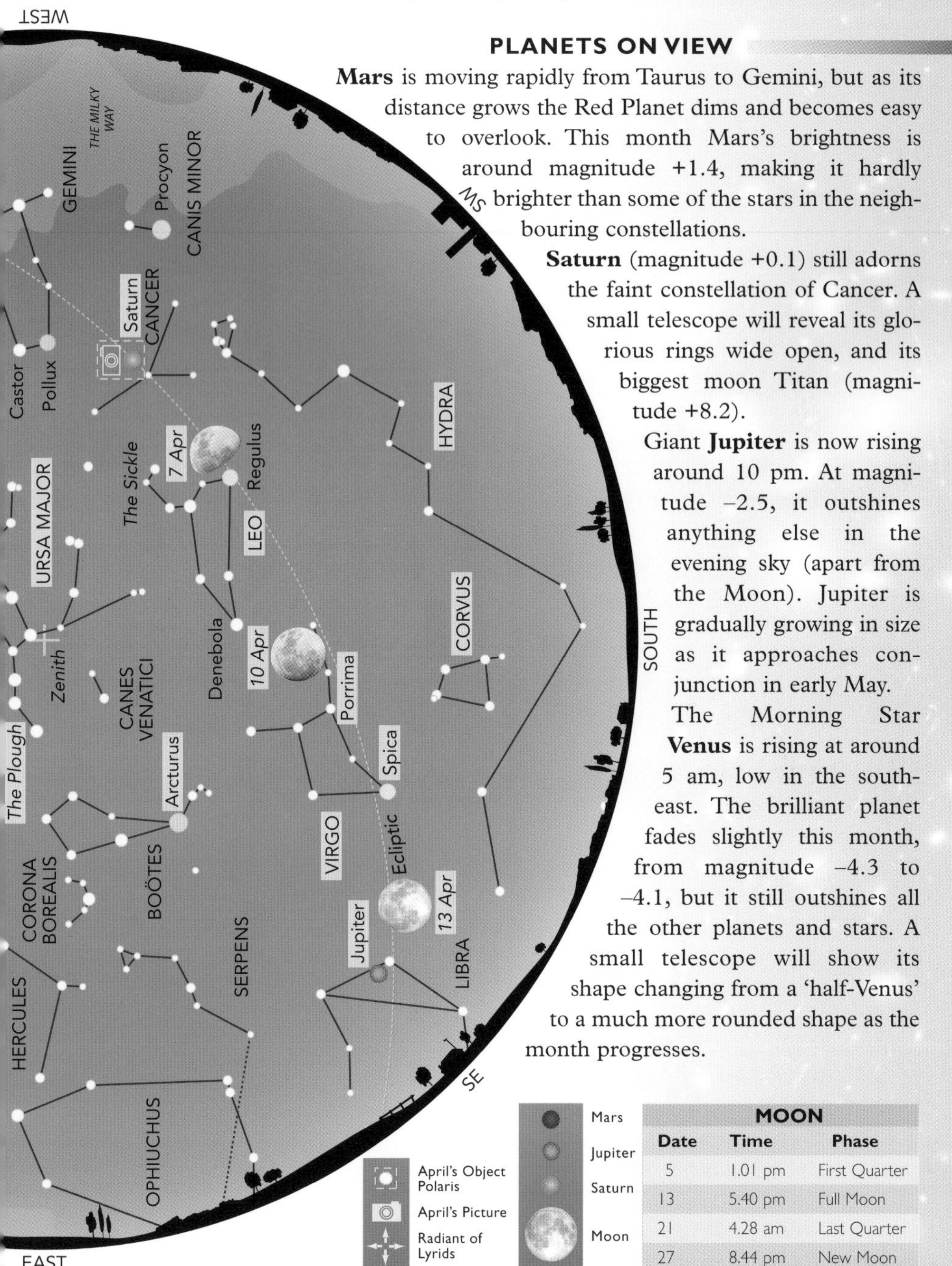

PLANETS ON VIEW

Mars is moving rapidly from Taurus to Gemini, but as its distance grows the Red Planet dims and becomes easy to overlook. This month Mars's brightness is around magnitude +1.4, making it hardly brighter than some of the stars in the neighbouring constellations.

Saturn (magnitude +0.1) still adorns the faint constellation of Cancer. A small telescope will reveal its glorious rings wide open, and its biggest moon Titan (magnitude +8.2).

Giant **Jupiter** is now rising around 10 pm. At magnitude –2.5, it outshines anything else in the evening sky (apart from the Moon). Jupiter is gradually growing in size as it approaches conjunction in early May.

The Morning Star **Venus** is rising at around 5 am, low in the southeast. The brilliant planet fades slightly this month, from magnitude –4.3 to –4.1, but it still outshines all the other planets and stars. A small telescope will show its shape changing from a 'half-Venus' to a much more rounded shape as the month progresses.

MOON

Date	Time	Phase
5	1.01 pm	First Quarter
13	5.40 pm	Full Moon
21	4.28 am	Last Quarter
27	8.44 pm	New Moon

Mercury, **Uranus** and **Neptune** are all lost in the morning twilight glow this month.

MOON

On 1 April, the crescent Moon is very close to the Pleiades star cluster: people in North America will see the Moon pass right in front of the Seven Sisters. The Moon lies close to Mars on 3 April, Saturn on 6 April and Jupiter on 14 and 15 April. The crescent Moon appears in the morning sky with brilliant Venus on 24 and 25 April.

SPECIAL EVENTS

22 April: It is the maximum of the **Lyrid** meteor shower, which – by perspective – appears to emanate from the constellation of **Lyra**. The shower, which consists of particles from a comet called Thatcher, is active between 19 and 25 April. Usually it generates a desultory 10 shooting stars per hour, but occasionally we have seen it offer more. This should be a good year for observing the Lyrids, as moonlight will not interfere.

APRIL'S OBJECT

The Pole Star – **Polaris**, in Ursa Minor – is a surprisingly shy animal, coming in at the modest magnitude of +2.1. But its importance throughout recent history centres on the fact that

Viewing tip
A medium-sized pair of binoculars can reveal Jupiter's four largest moons, but you have to use them carefully. Balance your elbows on a fence or a table to minimize wobble – the magnification of the binoculars only magnifies any shaking. You should see the moons strung out in a line either side of Jupiter's equator.

◀ A clever composite of Saturn and its moons, taken by Dave Tyler. The equipment used was a 280 mm Celestron telescope, and two webcams. A black-and-white ATK webcam (its sensitivity higher than a colour webcam) imaged the moons, while a colour ToUcam, used immediately afterwards, captured Saturn.

Earth's north pole points to Polaris, so we spin 'underneath' it. It remains stationary in the sky, and acts as a fixed point for both astronomy and navigation. You can find Polaris by following the two end stars of the Plough (as shown on the chart). Over a 26,000-year period, the Earth's axis swings around like an old-fashioned spinning top – a phenomenon called *precession* – so our 'pole stars' change with time. Polaris is currently about one degree off exact alignment with the north pole; it will be nearest to the 'above pole' position in AD 2100. Famous pole stars of the past include **Kochab** in Ursa Minor, which presided over the skies during the Trojan Wars of 1184 BC. In 14,000 years' time, brilliant **Vega**, in Lyra, will be our pole star.

APRIL'S PICTURE

Saturn still features in the skies of early spring, and this picture of the ring-world and its brightest moons captures the planet perfectly. It looks like a tiny model world, straight out of the pages of the *Eagle* comic. But appearances are deceptive: Saturn is huge, and its rings would stretch almost all the way from the Earth to the Moon.

APRIL'S TOPIC
Venus Express

In April, the brilliant planet Venus will be in the glare of the space spotlight, for the first time in over a decade. If the launch of Venus Express goes according to plan in autumn 2005, the planet will have a European space probe going into orbit around it this month. Although the Russians and Americans have very much made Venus their own planet, this is Europe's first attempt at a mission to our nearest world. The probe is being built by the same teams that were behind the hugely successful Mars Express, and the name is no misnomer – the probe travels to the planet in less than six months!

Once in orbit, Venus Express will study the planet's baffling atmosphere and its interaction with the planet's surface. Although Venus is so Earth-like in size, its atmosphere of carbon dioxide is 100 times heavier than ours, and has led to a runaway Greenhouse Effect, raising the temperature on the planet to 460°C. And Venus is still geologically active. Its oldest craters are no more than 500 million years old, suggesting that the planet is periodically 're-surfaced' – probably through massive volcanic eruptions. Venus Express will carefully monitor our neighbour-world to shed light on these mysteries.

This is an exciting month in astronomy, with pieces of a broken-up comet about to whiz past the Earth at very close range. The brightest fragment may be the most brilliant comet to grace the skies since Hale–Bopp in 1997. And it may be accompanied by a hail of shooting stars.

The cometary fireworks promise to upstage giant planet **Jupiter**, which is at its closest to Earth this month. **Mars** and **Saturn** also grace the evening skies, while Venus appears before dawn.

On the stellar front, bright orange **Arcturus** is high in the south, at the base of the kite-shaped constellation **Boötes**. To its right is the feline outline of **Leo** (the lion), with 1st-magnitude **Regulus** marking its heart. Below them both is the Y-shape of **Virgo** (the virgin), which boasts blue-white **Spica** as its brightest star.

MAY'S CONSTELLATION

For one of antiquity's superheroes, the celestial version of **Hercules** looks like a wimp. While Orion is all strutting masculinity, Hercules – one of the ancient Greeks' main legends, famous for his twelve labours – is but a poor reflection, and also upside-down. The two constellations are similar in shape, as Hercules also resembles the outline of a man, but its stars are faint and undistinguished.

It is, however, a fascinating constellation. Outside the rectangular main 'body' of the hero, to the south, is **Rasalgethi** – Hercules' head. With a girth 600 times that of the Sun, it is one of the biggest stars known. This distended object, close to the end of its life, flops and billows in its death throes. As a result, it varies in brightness, changing from 3rd to 4th magnitude over a period of about 90 days.

Hercules boasts one of the most spectacular sights in the northern night sky. Returning to the rectangular 'body', about

▼ The sky at 11 pm in the middle of May, with Moon-positions marked at intervals of three days. The positions of the stars are also correct for midnight

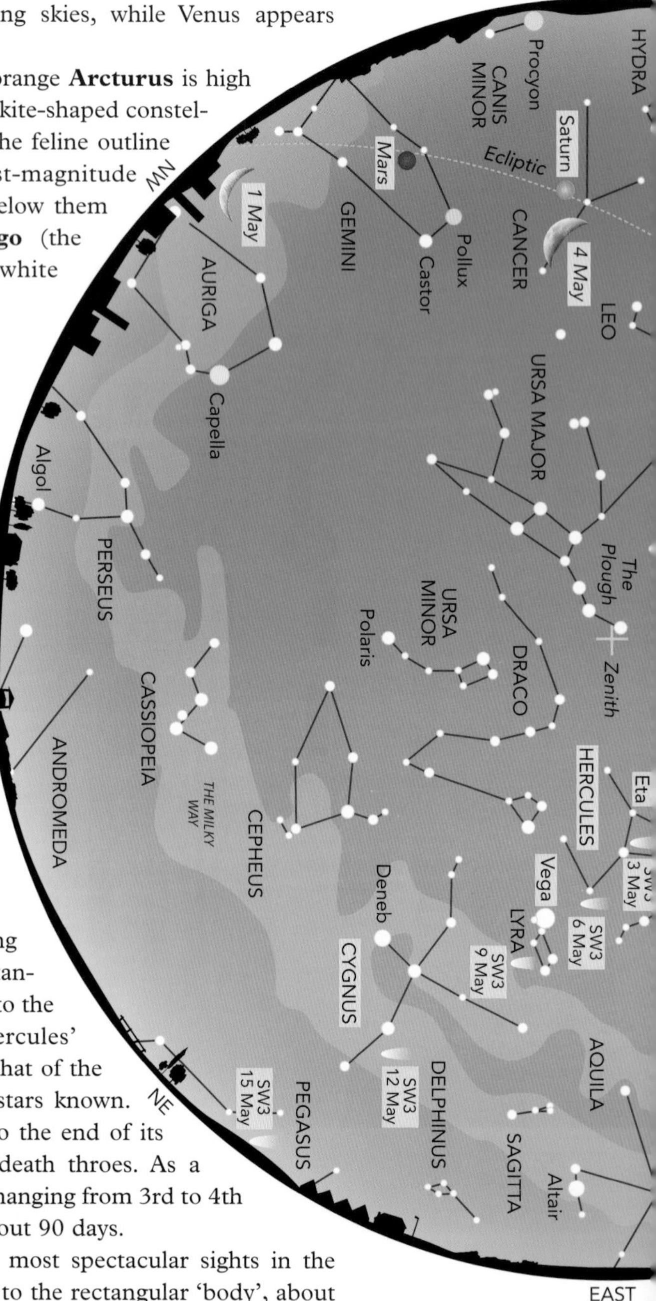

t the beginning of May, and 0 pm at the end of the month. he planets move slightly relative the stars during the course of he month.

a quarter of the way down from the top right-hand star (**Eta Herculis**) you will see a fuzzy patch. In a small telescope, **M13** – a globular cluster of almost a million stars – looks like a swarm of bees. There's a bonus this month, with Comet **Schwassmann–Wachmann 3** passing close by M13 on 1 and 2 May. More details are given in May's Topic.

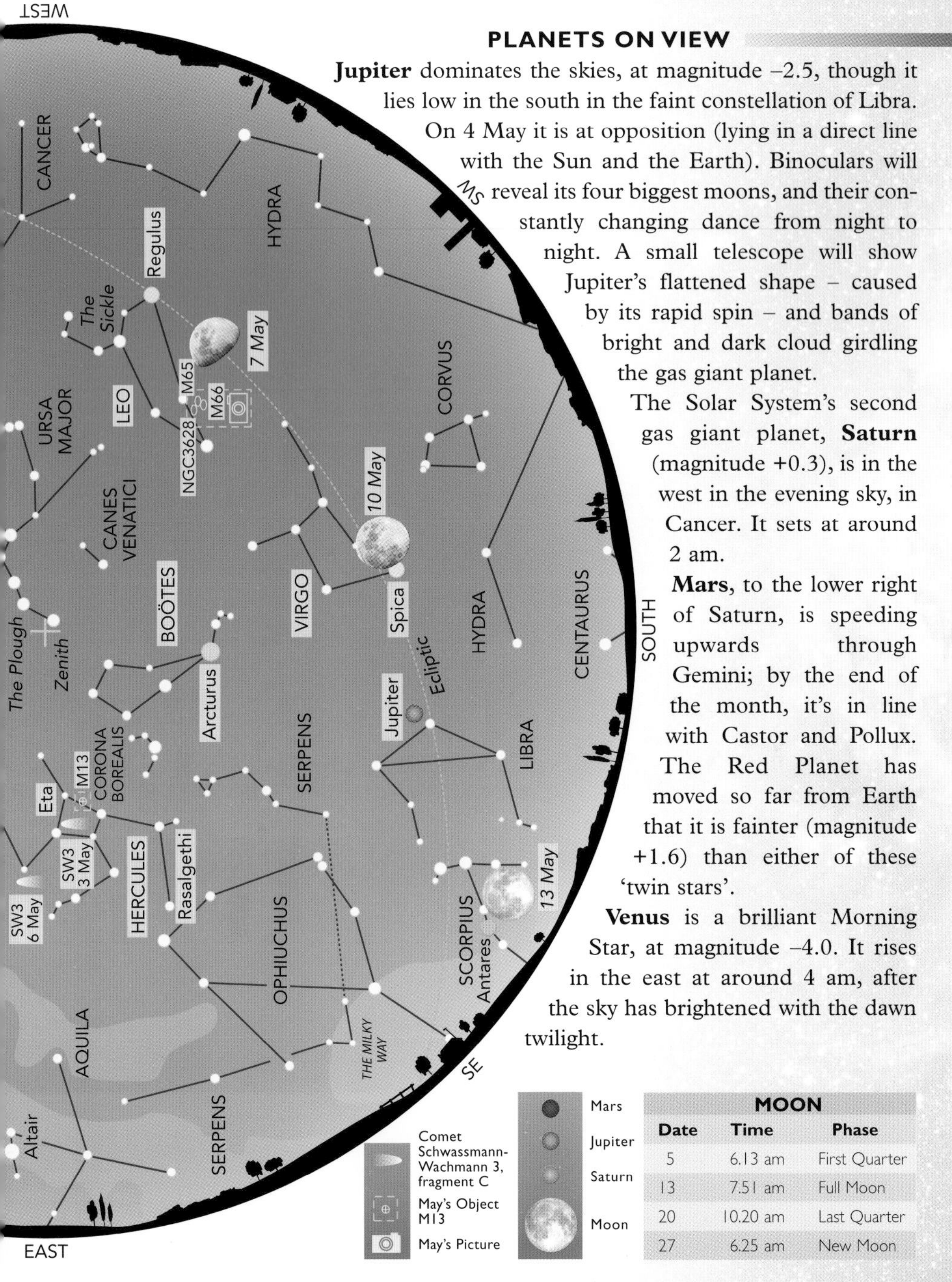

PLANETS ON VIEW

Jupiter dominates the skies, at magnitude –2.5, though it lies low in the south in the faint constellation of Libra. On 4 May it is at opposition (lying in a direct line with the Sun and the Earth). Binoculars will reveal its four biggest moons, and their constantly changing dance from night to night. A small telescope will show Jupiter's flattened shape – caused by its rapid spin – and bands of bright and dark cloud girdling the gas giant planet.

The Solar System's second gas giant planet, **Saturn** (magnitude +0.3), is in the west in the evening sky, in Cancer. It sets at around 2 am.

Mars, to the lower right of Saturn, is speeding upwards through Gemini; by the end of the month, it's in line with Castor and Pollux. The Red Planet has moved so far from Earth that it is fainter (magnitude +1.6) than either of these 'twin stars'.

Venus is a brilliant Morning Star, at magnitude –4.0. It rises in the east at around 4 am, after the sky has brightened with the dawn twilight.

MOON		
Date	**Time**	**Phase**
5	6.13 am	First Quarter
13	7.51 am	Full Moon
20	10.20 am	Last Quarter
27	6.25 am	New Moon

Neptune (magnitude +7.8), in Capricornus, is rising at around 2 am. **Mercury** and **Uranus** are too close to the Sun to be seen this month.

MOON

On 2 May, the crescent Moon lies above Mars and below the twin stars Castor and Pollux; the following night it passes Saturn. The gibbous Moon is very close to Virgo's brightest star, Spica, on the night of 10 May. On 11 and 12 May, the almost-Full Moon forms a striking pair with brilliant Jupiter. The morning of 24 May sees a stunning sight in the dawn twilight, with the narrow crescent Moon lying directly above the glorious Morning Star, Venus. On 30 May, the Moon is back in the evening sky, as a slim crescent in a line with Mars, Castor and Pollux; on the last night of the month, it is near Saturn.

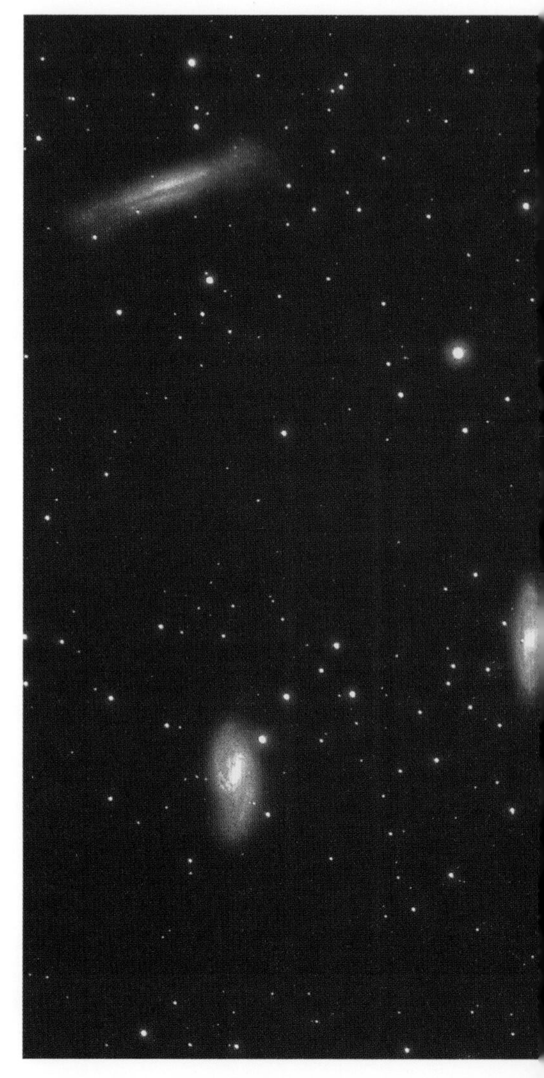

▼ *Taken in California by Michael Stecker, the lovely galaxies M66 (lower left) and M65 (lower right), along with NGC 3628, were imaged by a 155 mm f/7 refractor on Kodak PPF 35 mm film. The exposure time was 45 minutes.*

SPECIAL EVENTS

5 May: This is the maximum of the Eta Aquarid meteor shower. In early May you may notice a few extra shooting stars. Eta Aquarid meteors are tiny pieces of debris shed by Halley's Comet in its 76-year orbit of the Sun. The meteor shower's *radiant* – the point from which the meteors fan out as a result of perspective – is very low in the sky as seen from northern latitudes. People in the southern hemisphere may be lucky to see as many as 40 shooting stars per hour.

MAY'S OBJECT

At the darkest part of a May night, you may spot a faint fuzzy patch high in the south. Through binoculars, it appears as a gently glowing ball of light. A telescope reveals its true nature: a cluster of almost a million stars, swarming together in space.

This wonderful object is known as **M13** because it was the 13th entry in the catalogue of fuzzy objects recorded by the 18th-century French astronomer Charles Messier. We now classify M13 as a *globular cluster*, and it's one of 150 globulars that surround our Galaxy in its 'halo'.

Viewing tip

Unscrupulous mail-order 'gadget' catalogues often advertise small telescopes that boast huge magnifications. Beware! This is known as 'empty magnification' – blowing up an image that the lens or mirror simply doesn't have the light-grasp to get to grips with. A rule of thumb is to use a maximum magnification no greater than twice the diameter of your lens or mirror in millimetres. So if you have a 100-mm reflecting telescope, go no higher than 200X.

These great round balls of stars are among the oldest objects in our Milky Way Galaxy, dating back to its birth some 13 billion years ago. Researchers believe that these ancient collections of stars could have been the building blocks of our Galaxy.

In 1974, radio astronomers sent a message towards M13, hoping to inform the inhabitants of any planet there of our existence. M13 lies so far away, however, that we wouldn't receive a reply until AD 52,200!

MAY'S PICTURE

M66, **M65** and **NGC 3628** are three beautiful spiral galaxies in the constellation of **Leo**. They are members of a small group, rather like our own Local Group of galaxies. M66 and M65 can both be seen through a small telescope and are a lovely sight in a low-power eyepiece. NGC 3628 is also visible through a small telescope.

MAY'S TOPIC
Comet Schwassmann–Wachmann 3

Comets are notoriously unpredictable animals, Comet Schwassmann–Wachmann 3 particularly so! Discovered in Hamburg in May 1930 by the two German astronomers after whom it is named, the comet orbits the Sun roughly every five years.

In 1995, observers were surprised to find it undergoing a major outburst, reaching 5th magnitude (when expected to be at magnitude 12). Astronomers subsequently discovered that it had split into three fragments.

This month, the fragments are due to make a close approach to Earth, only 10 million km away – and one of the fragments may become visible to the naked eye, streaking across the sky at a rate of 4.5 degrees a day. The comet's path is marked on the chart, though its brightness is anyone's guess. Look towards the constellation of Cygnus on the night of 12/13 May when it may reach naked-eye visibility. And because the comet is coming so close, there's the possibility of a meteor shower, caused by cometary debris entering the Earth's atmosphere.

The midsummer month is the worst month for astronomy. The days are at their longest, and the nights at their shortest. The hours of darkness depend critically on latitude. In northern parts of Europe and Canada, it never truly gets dark. And north of the Arctic Circle (latitude 66.5°N), the Sun doesn't set at all. It just dips down towards the northern horizon, then moves up in the sky again, producing the phenomenon of the Midnight Sun.

At more temperate latitudes, the Sun is setting at its most northerly point along the horizon. This unique time of year was certainly picked out by our distant ancestors, who built great monuments like Stonehenge.

▼ The sky at 11 pm in the middle of June, with Moon-positions marked at intervals of three days. The positions of the stars are also correct for midnight

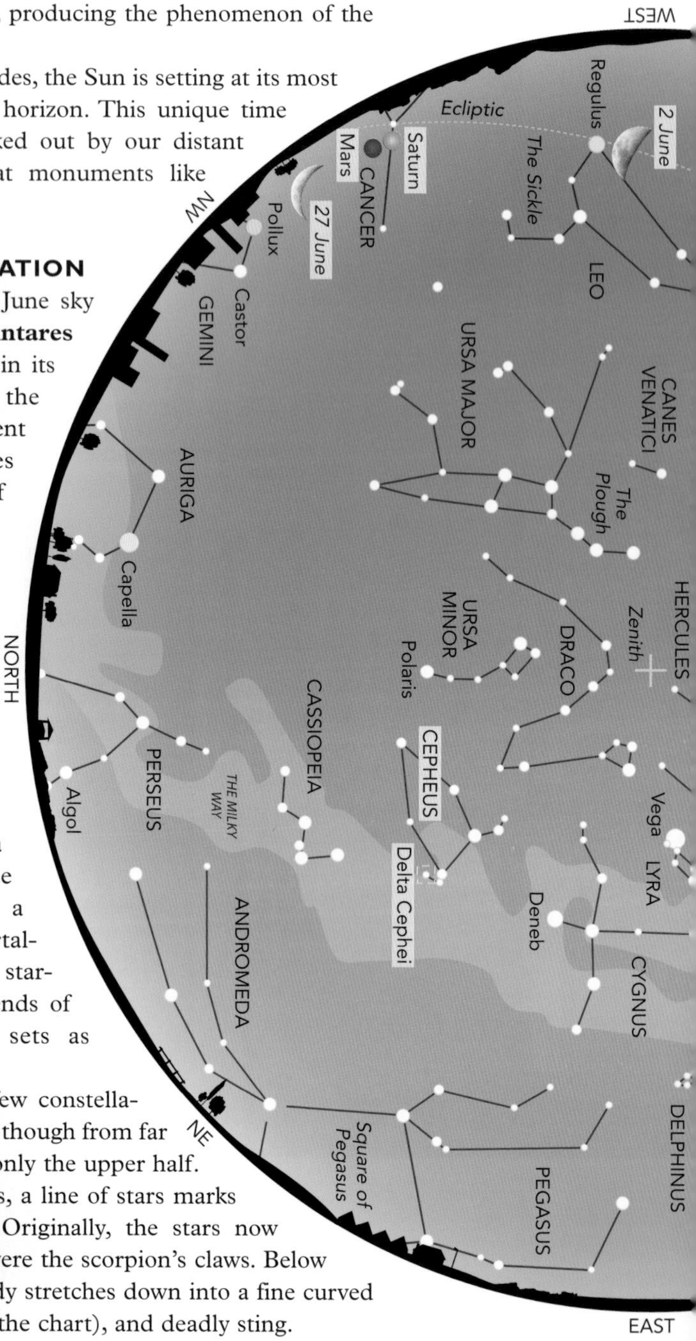

JUNE'S CONSTELLATION

In the deep south of the June sky lies a baleful red star – **Antares** ('the rival of Mars'); and in its ruddiness it surpasses the famed Red Planet. To ancient astronomers, Antares marked the heart of **Scorpius**, the celestial scorpion.

According to Greek mythology, this summer constellation is linked with the winter star-pattern Orion. The hunter boasted that he could kill every creature that lived. In retaliation, the Earth-goddess Gaia created a mighty scorpion that rose behind Orion, delivering a fatal sting. The gods immortalized these opponents as star-patterns, set at opposite ends of the sky so that Orion sets as Scorpius rises.

Scorpius is one of the few constellations to resemble its name, though from far northern latitudes we see only the upper half. To the top right of Antares, a line of stars marks the scorpion's forelimbs. Originally, the stars now called **Libra** (the scales) were the scorpion's claws. Below Antares, the scorpion's body stretches down into a fine curved tail (below the horizon on the chart), and deadly sting.

t the beginning of June, and 0 pm at the end of the month. he planets move slightly relative o the stars during the course of he month.

Scorpius has several lovely double stars. One of them is Antares, against whose strong red hue its faint companion looks greenish. To Antares' right, binoculars reveal the fuzzy patch of **M4**, a globular cluster made of tens of thousands of stars. It's one of the nearest of these giant clusters, a mere 7200 light years away.

The 'sting' contains three fine star clusters – **M6**, **M7** and NGC 6231, the last of which is below the chart's horizon. They can be seen with the naked eye.

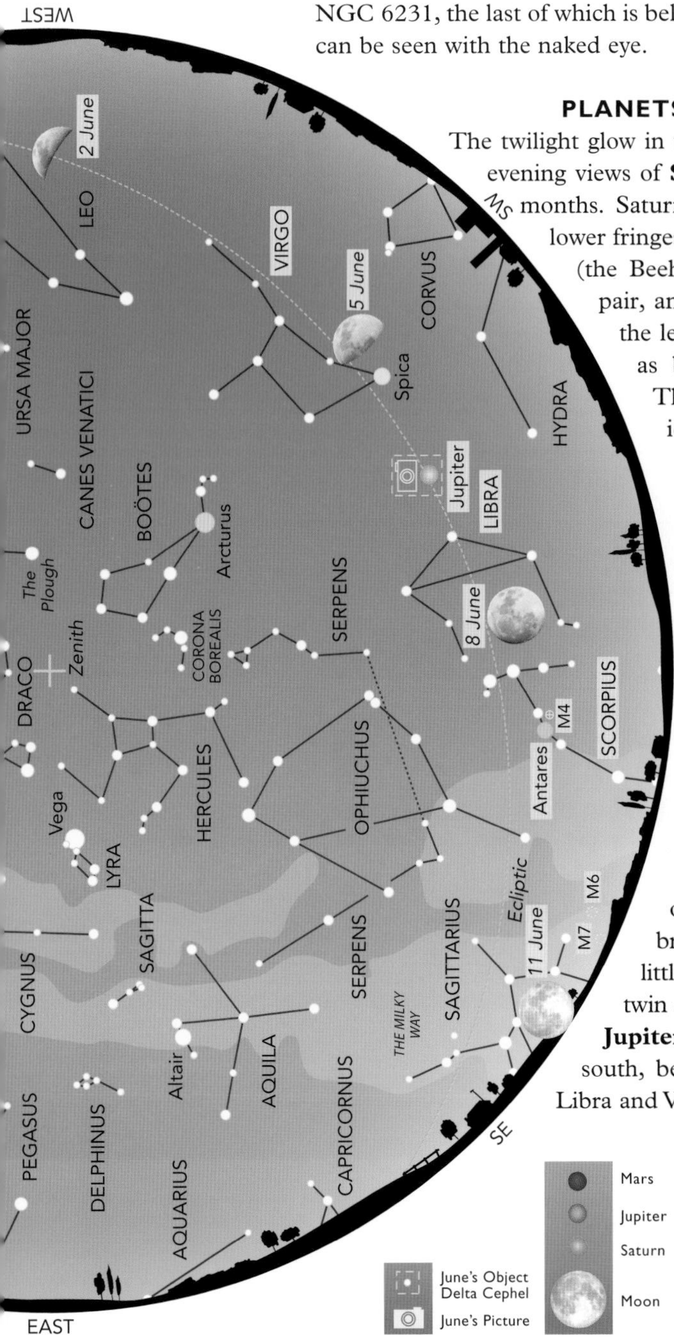

PLANETS ON VIEW

The twilight glow in the northwest offers the last evening views of **Saturn** and **Mars** for a few months. Saturn (magnitude +0.4), at the lower fringes of the star cluster Praesepe (the Beehive), is the brighter of the pair, and starts June as the one on the left. Mars is only one-quarter as bright, at magnitude +1.8. The Red Planet is moving rapidly to the upper left: on 14–16 June it passes in front of Praesepe, and makes its closest approach to Saturn on 17 June.

For much of June, **Mercury** hugs the northwestern horizon to the lower right of Mars immediately after sunset. It fades from magnitude –0.8 to +1.8 during the month. It reaches its greatest elongation from the Sun on 20 June, but is best seen one week earlier when it is brighter and the twilight glow a little fainter – look below the twin stars Castor and Pollux.

Jupiter shines brilliantly low in the south, between the constellations of Libra and Virgo. At magnitude –2.3, it's

Mars
Jupiter
Saturn
Moon

June's Object Delta Cephei
June's Picture

MOON		
Date	Time	Phase
4	0.06 am	First Quarter
11	7.03 pm	Full Moon
18	3.08 pm	Last Quarter
25	5.05 pm	New Moon

brighter than any of the stars, and – as with all the planets – it shines with a steady glow rather than twinkling like a star.

The Morning Star **Venus** is rising in the east less than two hours before the Sun, and shines at magnitude –3.9 in the dawn twilight.

Uranus shines at magnitude +5.8 in Aquarius, and **Neptune** is found (with a telescope) in Capricornus at magnitude +7.9.

Viewing tip
This is the month for maximum Sun-viewing, but you must be careful. Never use a telescope or binoculars to look at the Sun directly – it could blind you permanently. Fogged film is no safer, because it allows the Sun's infrared (heat) rays to get through. Eclipse goggles are safe (so long as they aren't scratched). The best way to observe the Sun is to project its image through binoculars or a telescope onto a white piece of card.

MOON

The Moon lies near brilliant Jupiter on 7 and 8 June. On 10 June, the almost-Full Moon lies low in the sky, under the red giant star Antares. The morning of 23 June sees the slender crescent Moon very low down in the Pleiades star cluster (you'll need binoculars to see the Seven Sisters in the twilight glow), with Morning Star Venus immediately below. On the evening of 27 June, there's a thin crescent low in the northwest after sunset, with the planets Saturn and Mars to its left, and Mercury right down on the horizon below; the Moon passes Mars the following evening.

SPECIAL EVENTS

11/12 June: The Full Moon tonight is the lowest for 18 years. From London it is only 9 degrees high; in Shetland you'll see the Moon literally skim the horizon.

21 June 1.26 pm: It is the Summer Solstice. The Sun reaches its most northerly point in the sky, so 21 June is Midsummer's Day, with the longest period of daylight (at the latitude of London, the Sun is up for 16 hours 39 minutes). Correspondingly, we have the shortest night.

JUNE'S OBJECT

At first glance, the variable star **Delta Cephei** – in the constellation representing King Cepheus – doesn't seem to merit special attention. It's a yellowish star of magnitude 4 – easily visible to the naked eye, but not prominent; a telescope reveals a companion star. But this type of star holds the key to the size of the Universe.

Check this star's brightness carefully over days and weeks, and you'll see that its brightness changes regularly, from +3.6 (brightest) to +4.3 (faintest), every 5 days 9 hours. This variation is a result of the star literally swelling and shrinking in size, from 32 to 35 times the Sun's diameter.

Astronomers have found that stars like this – *Cepheid variables* – show a link between their period of variation and their intrinsic luminosity. By observing the star's period and brightness, astronomers can work out a Cepheid's distance. With the Hubble Space Telescope, astronomers have measured Cepheids in the Virgo Cluster of galaxies, which lies 55 million light years away.

▲ *Using a 250 mm reflector at prime focus, Robin Scagell captured this image of Jupiter and its largest moons on ISO 400 film. He later digitally enhanced the satellites, as they are so much fainter than Jupiter itself.*

JUNE'S PICTURE

Jupiter is at its closest to the Earth now, and appears as a glorious sight in the sky – it's a creamy-coloured brilliant 'star'. Even in good binoculars, you can see the four biggest moons (there are over 60 in total). This view, taken through a small telescope, shows **Jupiter** with Io, Europa, Ganymede and Callisto.

JUNE'S TOPIC
The DAWN mission

By now, a daring NASA mission should be on its way to the asteroid belt. DAWN is due to blast off on 27 May towards two asteroids – Ceres (the bigger, at 930 km across) and Vesta (the brighter). DAWN won't reach Vesta until July 2010, but then it will orbit the rocky lump for a year. It will then depart for Ceres, arriving in August 2014.

The mission's aim is to investigate the formation and evolution of our Solar System. Asteroids and comets are the building blocks of planets, and give us clues to our origins – perhaps even the origin of life. The two asteroids couldn't be more different, which is why NASA has chosen them as targets. Ceres evolved with water present (there may even be frost or water vapour on its surface today); Vesta, on the other hand, originated in hot and violent circumstances. By studying the two asteroids, DAWN will be looking back to the earliest days of our Solar System.

This is a month of big sprawling faint constellations, studded with a few brilliant jewels of 1st-magnitude stars. Compared with the obvious star-patterns of winter, such as Orion and Gemini, the summer constellations take a little more time to work out.

The easiest way to navigate the summer sky is to use the brightest stars as signposts. Over in the west is the great Summer Triangle. Its corners are marked by **Vega** (top right), **Deneb** (top left) and **Altair** (bottom) – the brightest stars of their respective constellations **Lyra** (the lyre), **Cygnus** (the swan) and **Aquila** (the eagle).

Well to the right of Vega is the slightly brighter **Arcturus**, in kite-shaped **Boötes**. Between these two 1st-magnitude stars lie the hourglass-shaped constellation of **Hercules** and the distinctive circlet of stars making up **Corona Borealis** (the northern crown). Below **Hercules**, a large ring of stars forms the body of **Ophiuchus**, the serpent-bearer, who is wrestling with the snake **Serpens**. The snake's head pokes up from the stars of Ophiuchus towards Corona Borealis.

▼ The sky at 11 pm in the middle of July, with Moon-positions marked at intervals of three days. The positions of the stars are also correct for midnight

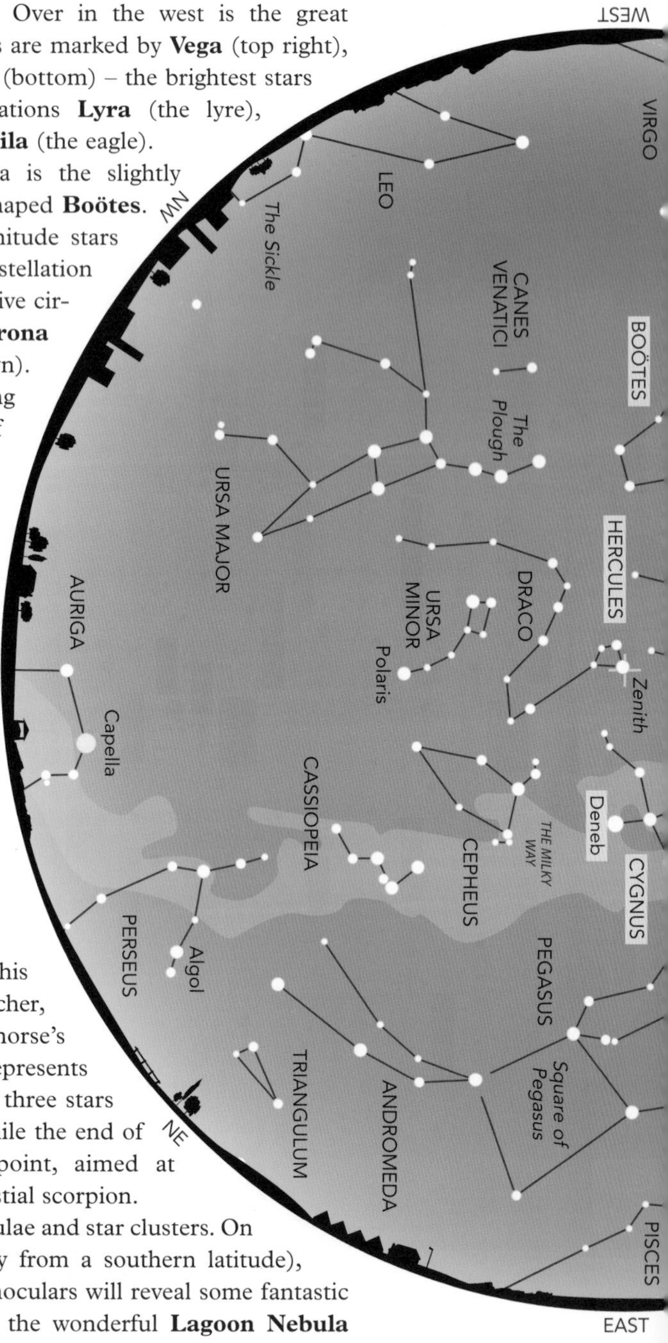

JULY'S CONSTELLATION

Low down in the south is **Sagittarius**, a constellation shaped rather like a teapot, with the handle lying to the left, and the spout to the right.

To the ancient Greeks, this star-pattern represented an archer, with a man's torso and a horse's body. The teapot's 'handle' represents his upper body, the curve of three stars to the right his bent bow, while the end of the spout is the arrow's point, aimed at **Scorpius**, the fearsome celestial scorpion.

Sagittarius is rich with nebulae and star clusters. On a clear night (and preferably from a southern latitude), sweeping Sagittarius with binoculars will reveal some fantastic sights. Above the spout lies the wonderful **Lagoon Nebula**

at the beginning of July, and 10 pm at the end of the month. The planets move slightly relative to the stars during the course of the month.

(M8) – visible to the naked eye on clear nights. This is a region where stars are being born. Between the 'teapot' and the neighbouring constellation Aquila is a bright patch of stars in the Milky Way (catalogued as **M24**). Higher up you'll spot another star-forming region, the **Omega Nebula** (M17).

On a very dark night you may notice a fuzzy patch, above and to the left of the teapot's lid. This is the globular cluster **M22**, a swarm of almost a million stars that lies around 10,000 light years away.

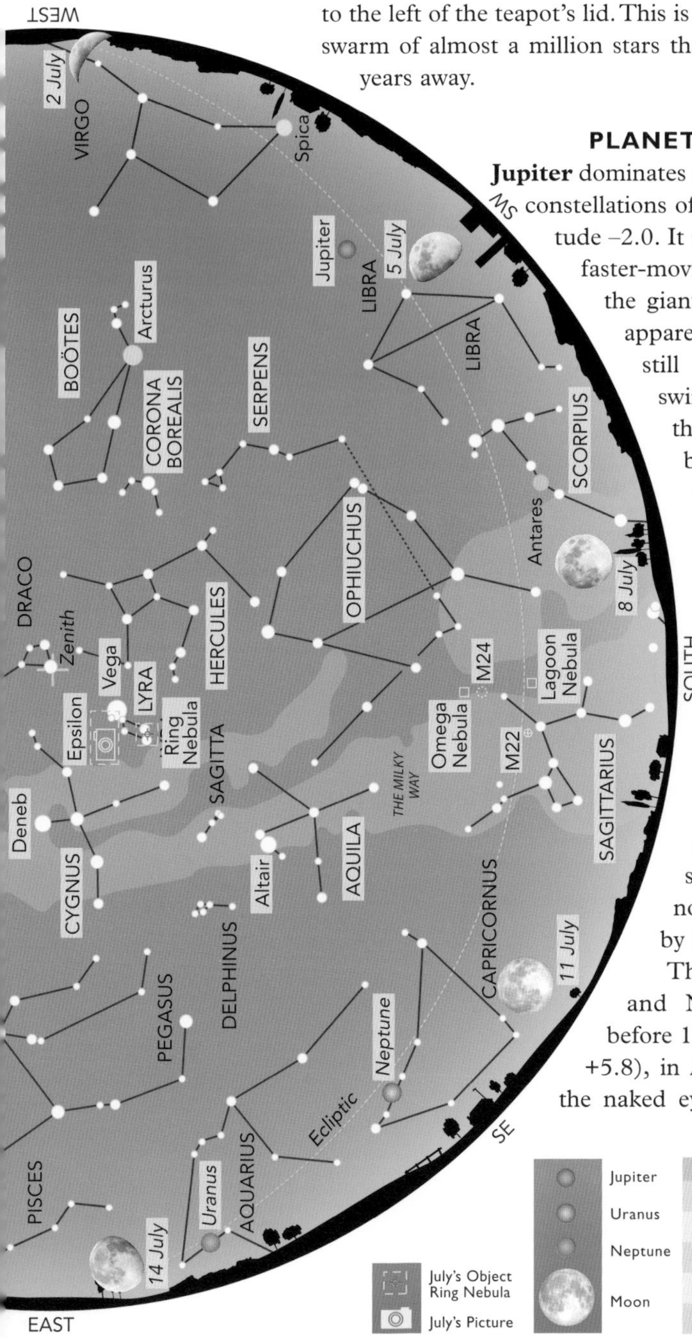

WEST

VIRGO
2 July
Spica
SW
Jupiter
5 July
LIBRA
LIBRA
Arcturus
BOÖTES
CORONA BOREALIS
SERPENS
SCORPIUS
Antares
OPHIUCHUS
8 July
DRACO
Zenith
Vega
LYRA
HERCULES
M24
Lagoon Nebula
SOUTH
Epsilon
Ring Nebula
SAGITTA
Omega Nebula
M22
THE MILKY WAY
SAGITTARIUS
Deneb
CYGNUS
Altair
AQUILA
CAPRICORNUS
11 July
DELPHINUS
PEGASUS
Neptune
Ecliptic
SE
PISCES
Uranus
AQUARIUS
14 July
EAST

PLANETS ON VIEW

Jupiter dominates the night sky, straddling the constellations of Virgo and Libra at magnitude –2.0. It sets at around 1 am. As the faster-moving Earth pulls away from the giant planet, Jupiter shrinks in apparent size. But a telescope will still reveal details in its ever swirling cloud patterns, and the nightly dance of its four biggest moons. On the evening of 2 July, Io, Europa, Ganymede and Callisto are all laid out in order on the right-hand side of Jupiter.

The Morning Star, **Venus**, is still low in the northeast before sunrise – rising at around 3 am. It shines at magnitude –3.9. As the Earth recedes, Venus looks ever smaller in the sky, and is now almost fully illuminated by the Sun.

The outer planets Uranus and Neptune are now rising before 11 pm. **Uranus** (magnitude +5.8), in Aquarius, is just visible to the naked eye, but is not easy to see

July's Object
Ring Nebula
July's Picture

Jupiter
Uranus
Neptune
Moon

MOON		
Date	Time	Phase
3	5.37 pm	First Quarter
11	4.02 am	Full Moon
17	8.12 pm	Last Quarter
25	5.31 am	New Moon

when so low in the sky. **Neptune** is in Capricornus, and at magnitude +7.8 you'll definitely need optical aid to spot it.

Mercury, **Mars** and **Saturn** are all too close to the Sun to be seen this month.

MOON

On 5 July, the Moon lies below the giant planet Jupiter. On the night of 14/15 July, the Moon passes immediately below the faint planet Uranus. Use this opportunity to locate the seventh planet by 'sweeping' upwards from the Moon with binoculars or a small telescope. Ignore the star immediately above the Moon (h Aquarii) to find Uranus about 1.5 moonwidths up. On 23 July, the narrow crescent Moon lies to the left of Venus in the pre-dawn sky.

SPECIAL EVENTS

3 July: Earth is at aphelion, its farthest point from the Sun. That's a distance of 152 million km.

JULY'S OBJECT

Tucked into the small constellation of Lyra (the lyre) – near the brilliant star Vega – lies a strange celestial sight. It was first spotted by French astronomer Antoine Darquier in 1779, who described it as 'a very dull nebula, but perfectly outlined; as large as Jupiter and looks like a fading planet'. Under higher

Viewing tip

In July you need a good, unobstructed view to the southern horizon in order to make out the summer constellations of Scorpius and Sagittarius. They never rise high in temperate latitudes, so make the best of a southerly view – especially over the sea – if you're away on holiday. A good southern horizon is also best for views of the planets, because they rise highest when they are in the south.

magnification, it appears as a bright ring of light with a dimmer centre. Hence its usual name of the **Ring Nebula** (M57).

You will see the Ring Nebula with even a small telescope, but it is so compact that a magnification of over 50× is needed in order to distinguish it from a star.

The Ring Nebula is a planetary nebula, the remains of a dying star; it is a cloud of gas lit up by the original star's incandescent core. For centuries, astronomers assumed that the Ring Nebula was a sphere of gas, but the Hubble Space Telescope found that it is in fact barrel-shaped. It looks like a ring to us only because we happen to view the barrel end-on. Aliens observing the Ring Nebula from another perspective would undoubtedly call it something different!

JULY'S PICTURE

Next to brilliant Vega lies one of the most unusual stars in the sky – **Epsilon Lyrae**. Through binoculars (or even with good eyesight) you can see that it is a double star, but a small telescope reveals that it's in fact a double-double. This beautiful image captures the colours of the four stars – in particular, the blue and yellow of the pair on the right.

JULY'S TOPIC
Stars

We think of the stars as being constant and unchanging, but a long look at the summer sky proves that they are anything but. Sweep Sagittarius with binoculars, and you'll find it studded with nebulae – gas clouds that are the nurseries of new stars. These fledgling stars grow up to make hot young stars like Vega, shining as a result of nuclear reactions in their searingly hot cores.

As stars age – and their hydrogen fuel starts to run out – they develop the problems of middle age. Look no further than orange Arcturus and baleful red Antares (in Scorpius) for stars that have swollen up and cooled down near the end of their lives. Stars like these will eventually puff off their distended atmospheres into space, leaving a brief-lived 'planetary nebula' (like the Ring Nebula in Lyra) surrounding the now-defunct core. The nebula will soon disperse, leaving the cooling core – a white dwarf – alone in space, destined to become a cold, black cinder.

◀ Damian Peach used a Vixen 80 mm refractor with a Barlow lens to take this image of the famous 'double-double' Epsilon Lyrae. The Barlow lens extended the focal length of the telescope to f/45, creating a bigger image. The picture was captured with a Fuji S2 Pro digital camera.

In this holiday month many of us are far from streetlights, and it's an ideal time for getting to know the sky. This year, brilliant Jupiter graces the heavens in the southwest during the early evening (setting at around 10.30 pm).

To the north lies the familiar seven-star pattern of **the Plough**, in **Ursa Major** (the great bear). The stars at its right-hand end point upwards to the Pole Star, **Polaris**. Continue the line to discover the M- or W-shape of **Cassiopeia** (a mythical queen).

High in the south is the **Summer Triangle**. To the east is the distinctive quadrilateral making up the **Square of Pegasus**, the body of the flying horse.

We can also expect some celestial fireworks, because August is the time of year for the most reliable meteor shower, the **Perseids**.

▼ The sky at 11 pm in the middle of August, with Moon-positions marked at intervals of three days. The positions of the stars are also correct for midnight

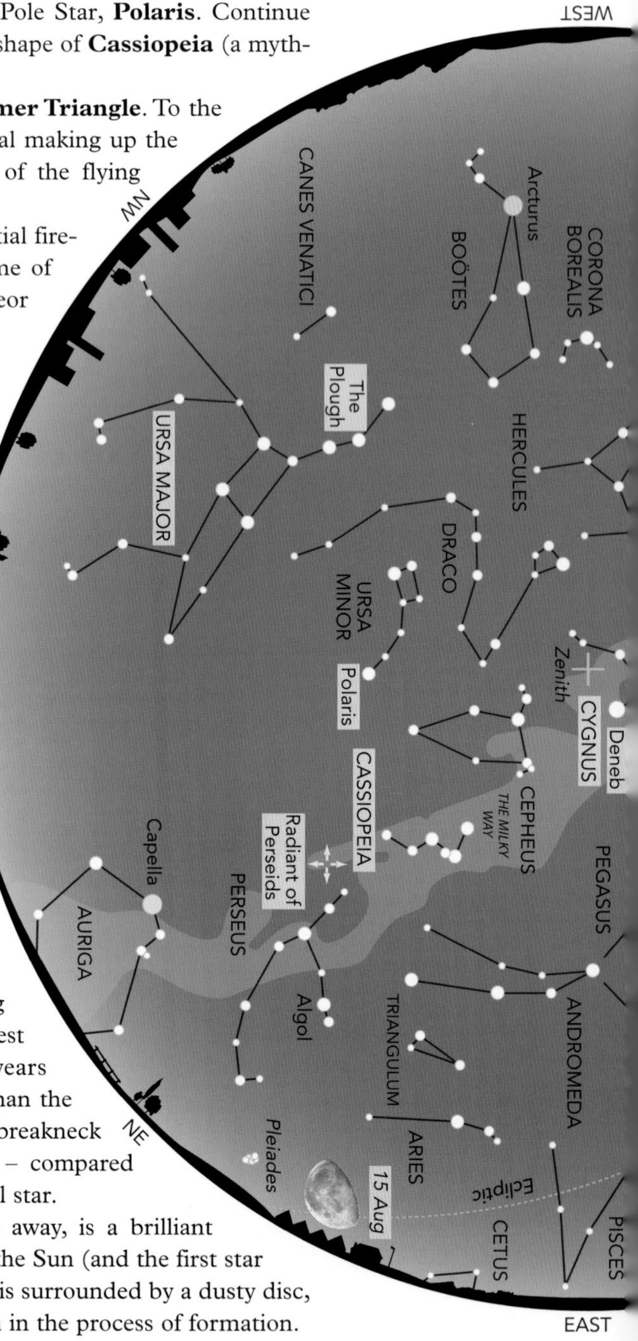

AUGUST'S CONSTELLATION

Strictly speaking a simple star-pattern rather than a constellation, the Summer Triangle is nevertheless a big part of the summer skies (around for most of autumn, too). The trio of **Vega**, **Deneb** and **Altair** – the brightest stars in their respective constellations of **Lyra**, **Cygnus** and **Aquila** – make a striking pattern almost overhead on August evenings.

The stars seem almost equally bright, but this is far from the case. Altair ('flying eagle') is one of the Sun's closest neighbours, nearly 17 light years away. About 10 times brighter than the Sun, it spins around at the breakneck speed of once every 6.5 hours – compared with around 30 days for our local star.

Vega, just over 25 light years away, is a brilliant white star nearly twice as hot as the Sun (and the first star to be photographed, in 1850). It is surrounded by a dusty disc, which may be a planetary system in the process of formation.

at the beginning of August, and 10 pm at the end of the month. The planets move slightly relative to the stars during the course of the month.

While Deneb ('tail' – of the swan) may appear to be the faintest of the three, the reality is quite different. It lies a staggering 3200 light years away, so to appear so bright it must be truly luminous. Deneb is over 200,000 times brighter than our Sun – one of the most brilliant stars known.

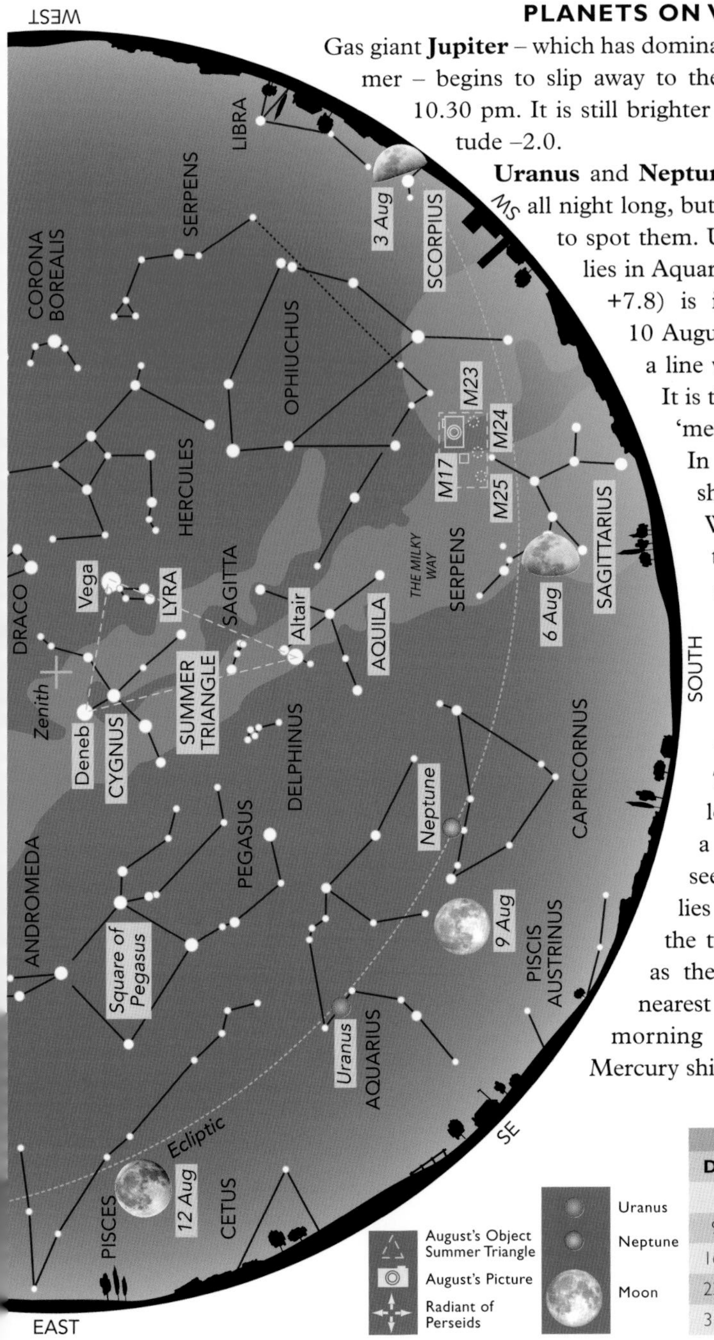

PLANETS ON VIEW

Gas giant **Jupiter** – which has dominated the night sky all summer – begins to slip away to the west, setting at around 10.30 pm. It is still brighter than the stars, at magnitude –2.0.

Uranus and **Neptune** are above the horizon all night long, but you will need optical aid to spot them. Uranus (magnitude +5.7) lies in Aquarius. Neptune (magnitude +7.8) is in Capricornus, and on 10 August reaches opposition (in a line with the Sun and Earth). It is then at its closest to us – a 'mere' 4340 million km away.

In the morning sky, **Venus** shines at magnitude –3.9. With sunrise coming later, the Morning Star is looking more resplendent in a darker sky.

Mercury is now a fainter morning star, reaching its greatest western elongation from the Sun on 7 August. But it is very low down and you'll need a clear eastern horizon to see it. On 1 August Mercury lies 10 degrees below Venus, the two planets moving closer as the days progress: they are nearest ($4\frac{1}{2}$ degrees apart) on the morning of 11 August, when Mercury shines at magnitude –0.4.

MOON		
Date	Time	Phase
2	9.46 am	First Quarter
9	11.54 am	Full Moon
16	2.51 am	Last Quarter
23	8.10 pm	New Moon
31	11.56 pm	First Quarter

▲ *This view of the star clouds and star clusters of Sagittarius was taken by Robin Scagell in Tenerife, where the lower latitude provides a better view of the constellation. He used a 50 mm Pentax lens at f/2 with Ektachrome 200 film; the exposure was five minutes.*

Mars is too close to the Sun to be seen this month. **Saturn** (magnitude +0.4) is also lost in the Sun's rays for most of August, but it creeps into the dawn twilight right at the month's end. On the mornings of 21 and 22 August it lies near Mercury, and early on 27 August is very close (½ degree) to brilliant Venus.

MOON

On the first night of August, the First Quarter Moon passes below brilliant Jupiter. The morning of 22 August sees an incredibly slender crescent Moon rising just before the Sun, in close company with the Morning Star, Venus; with binoculars and a clear horizon, you may see Mercury and Saturn lying below them. The Moon is near Jupiter again on the evening of 29 August.

SPECIAL EVENTS

12/13 August: This is the maximum of the **Perseid** meteor shower, and Perseid meteors will be seen for several nights around the time of maximum. The show will not be at its best this year, however, as bright moonlight will drown out the fainter meteors.

Viewing tip

Have a Perseids party! You don't need any optical equipment – in fact, telescopes and binoculars will restrict your view of the meteor shower. The ideal viewing equipment is your unaided eye, plus a sleeping bag and a lounger on the lawn. If you want to make measurements, a stopwatch and clock are good for timings, while a piece of string will help to measure the length of the meteor trail.

AUGUST'S OBJECTS

Not so much one object this month as millions of objects – the stars that make up our Milky Way. It was Galileo who, with his modest telescope, first ascertained the nature of the Milky Way, describing it as 'a congeries of stars'.

Sweep the band of light with modest binoculars this month, starting at **Cygnus**, then panning down to **Sagittarius** and **Scorpius**. What you are seeing are the more distant stars of our spiral galaxy. Because we live in the flat disc of the Galaxy, perspective makes these stars appear to congregate into a band. Looking towards Sagittarius, the stars look thickest – that's because you're seeing the dense nuclear bulge of the Milky Way. The dark patches you'll notice are not voids of stars: they're huge clouds of dark dust and gas poised to form new generations of stars and planets.

AUGUST'S PICTURE

Sagittarius, in the direction of the densely packed centre of our Galaxy, is home to many nebulae and star clusters. **M23** (right) and **M25** (left) are bright groups of young, hot stars visible to the naked eye; **M24** (centre) is not a true star cluster, but a bright, unobscured part of the Milky Way. At the top of the picture is **M17**, a nebula in which stars have just been born.

AUGUST'S TOPIC
Shooting stars

Many people report that they see loads of shooting stars on their summer holidays, and there's no mystery as to why. Between 8 and 12 August, Earth's orbit intersects a stream of debris from Comet Swift–Tuttle, which pours into the atmosphere at speeds of 210,000 km/h and burns up. This forms the Perseids – the most reliable meteor shower of the year. Because of perspective, the meteors all appear to diverge from the same part of the sky – the meteor's *radiant*, which lies in the constellation Perseus.

There's no chance of being struck by a meteor. These tiny particles of dust (less than a centimetre across) burn up around 60 km above the Earth's surface. Look upon the shower as a celestial fireworks display!

This is a pretty poor month for planets. As you can see on the chart, there are no planets visible to the naked eye in the evening sky (apart from borderline Uranus). The starry skies are fielding only a few bright objects too, notably the Summer Triangle – **Deneb**, **Vega** and **Altair**.

From a really dark site, what will catch your eye most this month is the glorious glowing band of the Milky Way. In September, it arches right overhead, from **Sagittarius** in the southwest, up through the cross of **Cygnus** and then down again through W-shaped **Cassiopeia** to the northeast horizon.

▼ *The sky at 11 pm in the middle of September, with Moon-positions marked at intervals of three days. The positions of the stars are also correct for midnight*

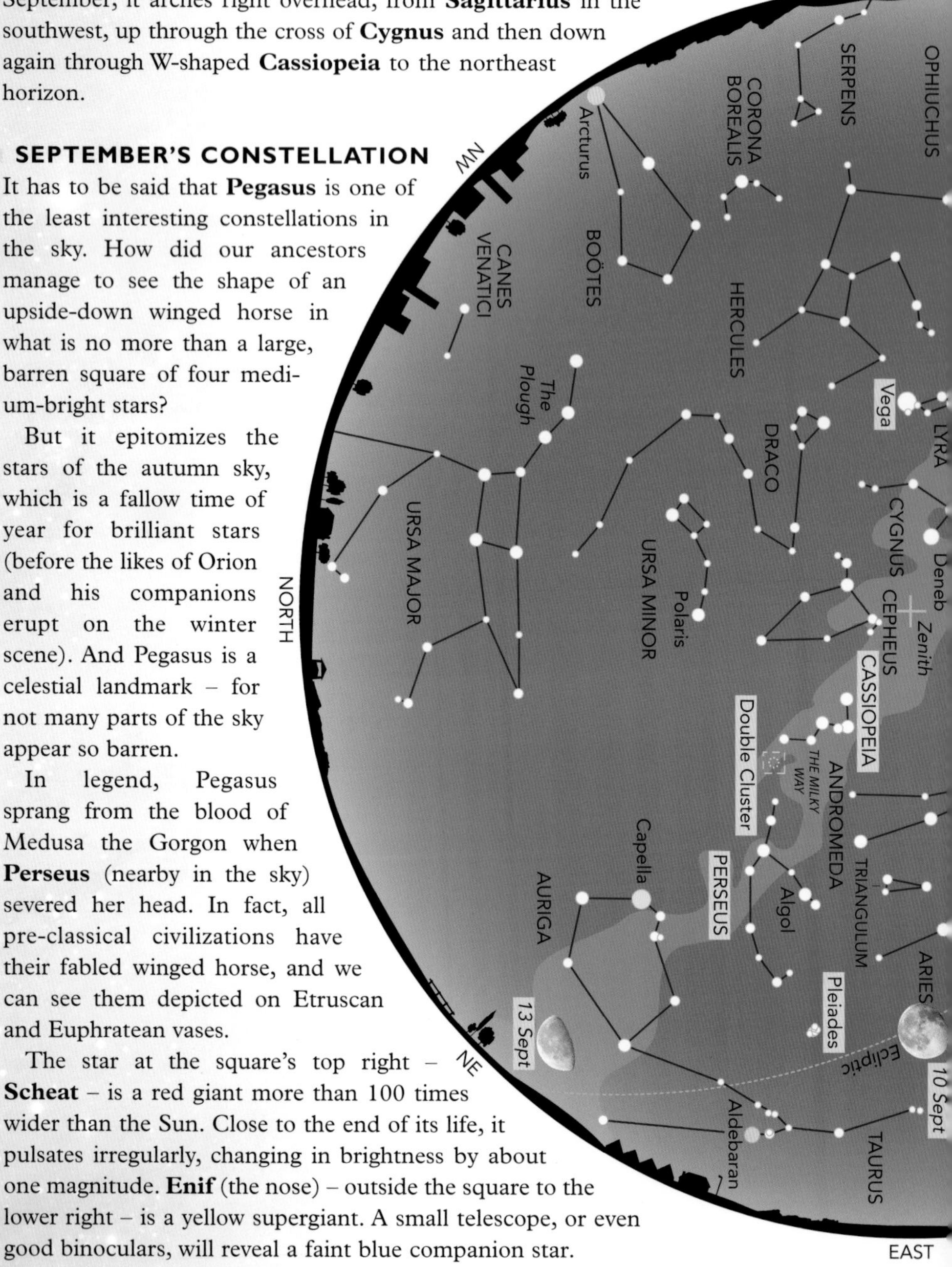

SEPTEMBER'S CONSTELLATION

It has to be said that **Pegasus** is one of the least interesting constellations in the sky. How did our ancestors manage to see the shape of an upside-down winged horse in what is no more than a large, barren square of four medium-bright stars?

But it epitomizes the stars of the autumn sky, which is a fallow time of year for brilliant stars (before the likes of Orion and his companions erupt on the winter scene). And Pegasus is a celestial landmark – for not many parts of the sky appear so barren.

In legend, Pegasus sprang from the blood of Medusa the Gorgon when **Perseus** (nearby in the sky) severed her head. In fact, all pre-classical civilizations have their fabled winged horse, and we can see them depicted on Etruscan and Euphratean vases.

The star at the square's top right – **Scheat** – is a red giant more than 100 times wider than the Sun. Close to the end of its life, it pulsates irregularly, changing in brightness by about one magnitude. **Enif** (the nose) – outside the square to the lower right – is a yellow supergiant. A small telescope, or even good binoculars, will reveal a faint blue companion star.

at the beginning of September, and 10 pm at the end of the month. The planets move slightly relative to the stars during the course of the month.

Just next to Enif – and Pegasus' best-kept secret – is the beautiful globular cluster **M15**. You'll need a telescope to view it. M15 is around 50,000 light years away, and contains about 200,000 stars.

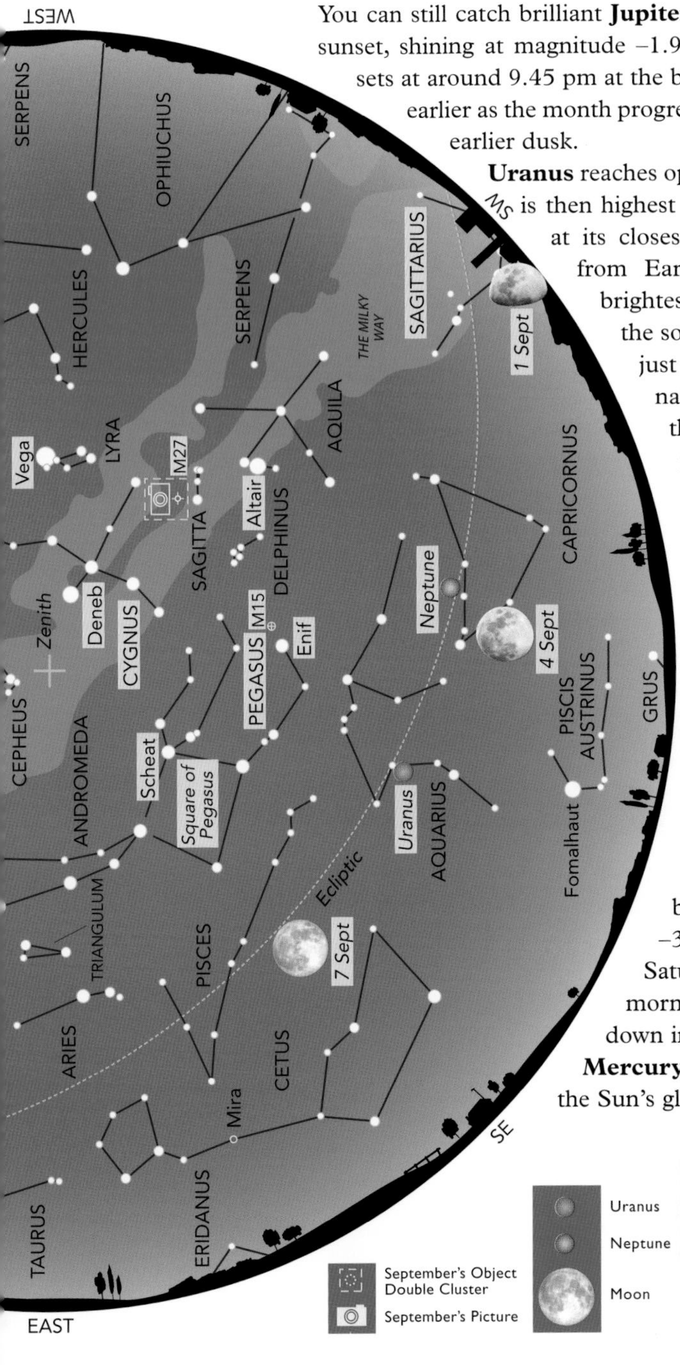

PLANETS ON VIEW

You can still catch brilliant **Jupiter** low in the southwest after sunset, shining at magnitude –1.9 in Libra. The giant planet sets at around 9.45 pm at the beginning of September, and earlier as the month progresses – keeping pace with the earlier dusk.

Uranus reaches opposition on 5 September. It is then highest in the sky at midnight, and at its closest – 2850 million km away from Earth. Uranus is also at its brightest, at magnitude +5.7. If in the southern hemisphere, you can just make out Uranus with the naked eye; but from the UK the planet is lost in atmospheric haze, and requires binoculars at least to spot it. More distant **Neptune** (magnitude +7.9) lies in Capricornus and requires a telescope.

Saturn (magnitude +0.5) is a morning object in Leo, rising at around 3.30 am. At the beginning of September you'll find it low down in the morning twilight to the east, close to the much brighter **Venus** (magnitude –3.9). During the month, Saturn rises ever higher in the morning sky, while Venus drops down into the dawn twilight.

Mercury and **Mars** are both lost in the Sun's glare this month.

WEST
SERPENS
OPHIUCHUS
HERCULES
SERPENS
THE MILKY WAY
SAGITTARIUS
SW
1 Sept
AQUILA
Vega
LYRA
M27
Altair
DELPHINUS
SAGITTA
CAPRICORNUS
Neptune
Deneb
Zenith
CYGNUS
M15
PEGASUS
Enif
4 Sept
PISCIS AUSTRINUS
GRUS
SOUTH
CEPHEUS
ANDROMEDA
Scheat
Square of Pegasus
Uranus
AQUARIUS
Fomalhaut
Ecliptic
TRIANGULUM
PISCES
7 Sept
ARIES
CETUS
Mira
SE
TAURUS
ERIDANUS
EAST

September's Object
Double Cluster
September's Picture

Uranus
Neptune
Moon

MOON		
Date	Time	Phase
7	7.42 pm	Full Moon
14	12.15 pm	Last Quarter
22	12.45 pm	New Moon
30	12.04 pm	First Quarter

MOON

As the Moon rises (at around 9 pm) on 12 September, you'll see it lying in the middle of the **Pleiades** star cluster. Over the next hour, it will occult (hide and then uncover) three of the brightest of the Seven Sisters – but you will need a clear north-east horizon and binoculars or a telescope to witness this event. On the morning of 19 September, the waning crescent Moon lies near to Saturn. The thinnest of lunar crescents appears near brilliant Venus in the dawn skies of 21 September.

SPECIAL EVENTS

7 September: There is a partial eclipse of the Moon. Looking carefully at the Moon as it rises, at around 7.30 pm, it will appear as though a chunk has been taken out of the top of it. This partial eclipse is not particularly spectacular. A maximum of 19% of the Moon is in the Earth's shadow; and the eclipse is over by 8.40 pm, before the sky is even totally dark.

22 September: There is an annular eclipse of the Sun. The New Moon moves over the face of the Sun, but because the Moon is at the far point of its orbit a ring of sunlight surrounds the Moon's silhouette. This annular eclipse is visible only from the South Atlantic; a partial eclipse will be visible from regions in South America and western Africa, but from the UK it is not visible at all.

23 September 05.03 am: It is the Autumn Equinox. The Sun is over the equator as it heads southwards in the sky, and day and night are equal in length.

Viewing tip

Try to observe your favourite objects when they're well clear of the horizon. When you look low down, you are seeing through a large thickness of the atmosphere, which is always shifting and turbulent. It's like trying to observe the outside world from the bottom of a swimming pool! This turbulence makes the stars appear to twinkle. Low-down planets also twinkle, but because they subtend tiny discs the twinkling is less marked.

SEPTEMBER'S OBJECTS

We have two objects this time: the glorious '**Double Cluster**' in **Perseus**. These near-twin clusters of young stars – each covering an area bigger than the Full Moon – are visible to the naked eye and are a gorgeous sight in binoculars. They are packed with pretty young blue stars, and are both around 7000 light years away. They are known as h and Chi Persei, with h being the older cluster at 5.6 million years, while Chi is a mere 3.2 million years old. Each contains around 200 stars, and is part of what's known as the Perseus OB1 Association – a loose group of bright, hot stars that were born at roughly the same time. Associations and star clusters allow astronomers to monitor stars that are the same age but have different masses – and the comparison helps researchers understand how stars evolve.

SEPTEMBER'S PICTURE

The Dumbbell Nebula – **M27** – is the 'star' of the tiny constellation Vulpecula, lying just above the brighter constellation Sagitta. One of the brightest planetary nebulae visible, M27 lies close to another planetary in the sky, the Ring Nebula in Lyra. Both are old stars that have jettisoned their atmospheres into space, revealing the hot core that once powered them. The glowing gas will dissipate in a mere few thousand years.

SEPTEMBER'S TOPIC
Constellations

Pegasus, our constellation of the month, highlights humankind's obsession to 'join up the dots' in the sky, and weave stories around them. But why did they do this? With the stars on view changing during the year (as the Earth moves around the Sun), the named constellations acted as an *aide-mémoire* to our position in the annual cycle – something of particular use to farming communities in ancient times.

The stars were also a great aid to navigation at sea. In fact, scholars believe that the Greek astronomers 'mapped' their legends onto the sky, so that sailors crossing the Mediterranean would recognize certain constellations through their knowledge of the traditional stories.

Not all the world saw the sky through western eyes. The Chinese divided up the sky into a plethora of tiny constellations, containing three or four stars apiece. The Australian Aborigines, in their dark deserts, were so overwhelmed with stars that they made constellations out of the dark places where they couldn't see any stars!

◀ The Dumbbell Nebula (M27) – named after its double appearance – was captured here by Damian Peach. He used a 127 mm Maksutov telescope, with a one-minute exposure on a Fuji Pro digital camera.

The evenings are now drawing in, providing splendid views of the night sky at relatively social hours – especially at the end of the month, when the clocks go back and the summer's extra hour of evening light is lost.

Well on display is the great **Square of Pegasus**, high in the south. Below the celestial flying horse are some constellations of a distinctly aqueous nature: **Pisces** (the fishes), **Cetus** (the sea-monster), **Aquarius** (the water-carrier) and that strange creature **Capricornus**, which is depicted in old star charts as half-goat and half-fish. Way down to the south is **Piscis Austrinus** (the southern fish), with the lovely star **Fomalhaut** (1st magnitude), which barely peeks above the horizon as seen from the UK.

▼ *The sky at 11 pm in the middle of October, with Moon-positions marked at intervals of three days. The positions of the stars are also correct for midnight at the*

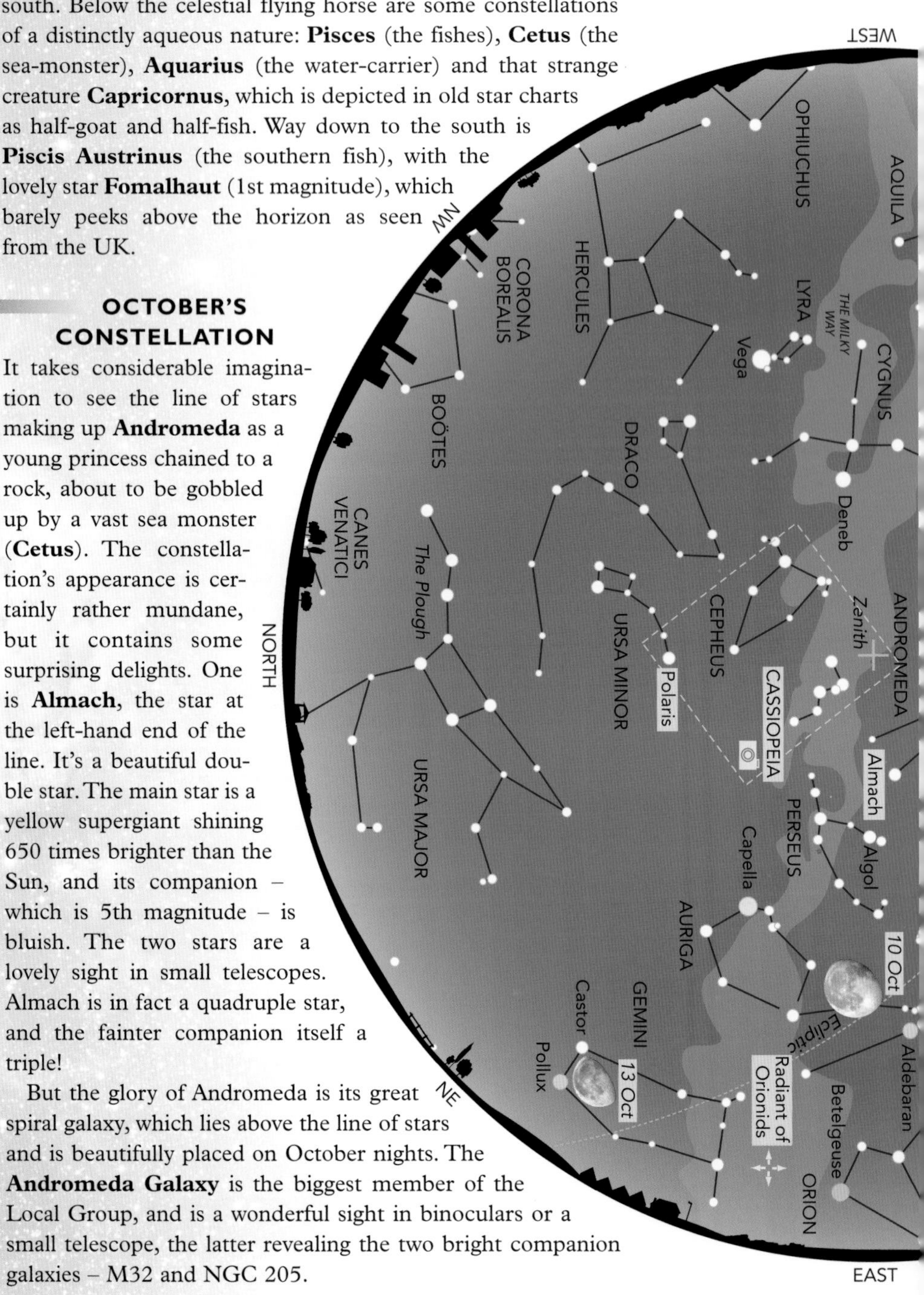

OCTOBER'S CONSTELLATION

It takes considerable imagination to see the line of stars making up **Andromeda** as a young princess chained to a rock, about to be gobbled up by a vast sea monster (**Cetus**). The constellation's appearance is certainly rather mundane, but it contains some surprising delights. One is **Almach**, the star at the left-hand end of the line. It's a beautiful double star. The main star is a yellow supergiant shining 650 times brighter than the Sun, and its companion – which is 5th magnitude – is bluish. The two stars are a lovely sight in small telescopes. Almach is in fact a quadruple star, and the fainter companion itself a triple!

But the glory of Andromeda is its great spiral galaxy, which lies above the line of stars and is beautifully placed on October nights. The **Andromeda Galaxy** is the biggest member of the Local Group, and is a wonderful sight in binoculars or a small telescope, the latter revealing the two bright companion galaxies – M32 and NGC 205.

beginning of October, and 9 pm at the end of the month (after the end of BST). The planets move slightly relative to the stars during the course of the month.

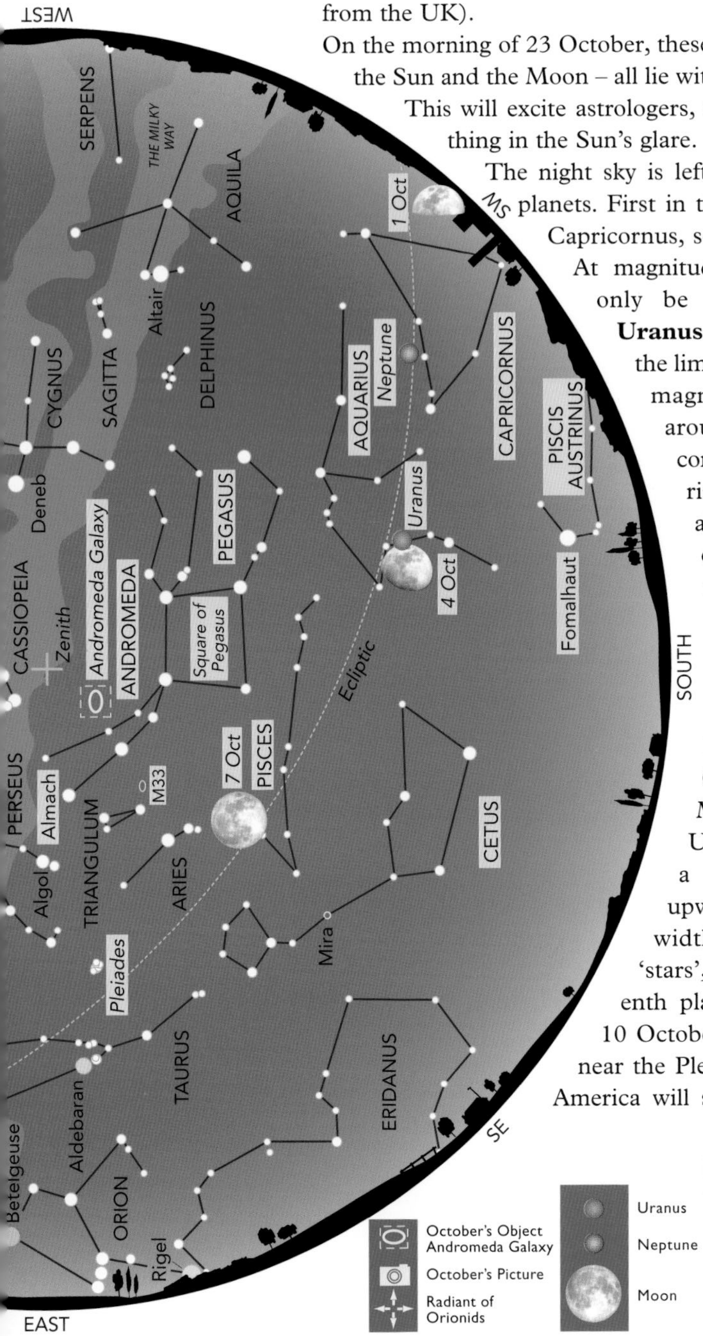

PLANETS ON VIEW

October is even worse for planets than September! For the first few days of the month, you may catch **Jupiter** very low in the southwest after sunset. Thereafter it is hidden in the Sun's brilliant glow, along with **Venus**, **Mars** and **Mercury** (which is at greatest elongation on 17 October, but still not easy to spot from the UK).

On the morning of 23 October, these four planets – along with the Sun and the Moon – all lie within 23 degrees on the sky. This will excite astrologers, though we won't see anything in the Sun's glare.

The night sky is left to the three outer giant planets. First in the parade is **Neptune**, in Capricornus, setting at around 1.30 am. At magnitude +7.9, the planet will only be seen with optical aid. **Uranus** in Aquarius follows – on the limit of naked-eye visibility at magnitude +5.8. It sets at around 3.30 am. Finally comes ringed planet **Saturn**, rising in the east in Leo, at about 2 am. This is the only planet obvious to the naked eye in October, shining at magnitude +0.5.

MOON

During the night of 4/5 October, the almost-Full Moon passes very near Uranus: with binoculars or a small telescope, sweep upwards about two moon-widths and you'll see two 'stars', the fainter being the seventh planet. On the morning of 10 October, the gibbous Moon lies near the Pleiades: observers in North America will see it occult some of the

MOON		
Date	Time	Phase
7	4.13 am	Full Moon
14	1.25 am	Last Quarter
22	6.14 am	New Moon
29	9.25 pm	First Quarter

Seven Sisters. On the morning of 16 and 17 October, the waning Moon lies near Saturn.

SPECIAL EVENTS

21 October: It is the maximum of the **Orionid** meteor shower. These meteors are debris from Halley's Comet, burning up in Earth's atmosphere, and they appear to spray out from the constellation Orion. This is a good year for observing the meteors, as moonlight will not interfere.
29 October, 2 am: British Summer Time ends. Clocks go backwards by one hour.

OCTOBER'S OBJECT

The **Andromeda Galaxy**, M31, looking from the UK like a fuzzy dinner plate on its side in the sky, is for most of us the most distant object visible to the naked eye. (The Triangulum Galaxy, **M33**, is fractionally further away but only tentatively visible from dark desert locations.)

Recent data from the Hipparcos satellite show the galaxy to be 2.9 million light years away, making it even more remote than originally thought. Despite its distance, Andromeda is so

◀ From Mount Fuji – at an altitude of 2400 m – Yoji Hirose captured this image of Cassiopeia with a 55 mm lens attached to a Mamiya M645 camera. He was using Fujichrome 400 film, and the exposure lasted 20 minutes.

vast – one of the biggest spiral galaxies known – that it appears nearly four times as big as our Full Moon (although the sky will seldom be clear enough to allow the faint outer spiral arms to be seen).

When you look at this amazing object through binoculars or a small telescope, you have to pinch yourself to remember that you're looking at 400 billion stars. It's a great shame that it is presented to us at such a steep angle that more of its amazing spiral structure can't be observed.

OCTOBER'S PICTURE

W-shaped **Cassiopeia** is always a feature of our northern skies. It's so close to the pole that it never sets, and forever swings around **Polaris** (top left) as the Earth turns. The ancient Greeks saw the stars as making up a queen in a chair; the Chinese saw them as a path through the mountains, and a charioteer with his horses.

OCTOBER'S TOPIC
Galaxies

With the Andromeda Galaxy riding high, and the stars of our own Galaxy all around us, it's time to take a look at these billions of star-cities that populate the Universe. Both the Milky Way and the Andromeda Galaxy are 'spirals' – they're rich in gas and dust poised to make new generations of stars, and are adorned with beautiful curving arms made of young, hot stars.

Irregular galaxies have a similar mix of ingredients but are too small to 'grow' arms. If you visit the southern hemisphere, you can see our two irregular companions, the Large and Small Magellanic Clouds, shining brightly in the sky.

Ellipticals make up the third category of galaxy. These range from the very small to the truly gargantuan, some with trillions of stars. But most of those stars are old and red; there's very little by way of building materials in these galaxies to make new stars.

A few galaxies are violent, with brilliant jets of gas shooting out from the vicinity of a central black hole at speeds close to the velocity of light.

Viewing tip

Although the Andromeda Galaxy is the farthest object easily visible to the unaided eye, it is large and extended, and not that easy to spot. The trick is to memorize the star patterns in the constellation Andromeda and look slightly to the side of where you expect the galaxy to be. This technique – called 'averted vision' – causes the image to fall on a part of the eye's retina that is more light-sensitive than the central part, which is designed to see fine detail.

It is truly autumn now, and that's reinforced by what can be seen in the sky above us. The faint, sprawling constellations of **Pegasus**, **Andromeda**, **Pisces** and **Cetus** make the heavens look lacklustre – a reflection of the autumnal landscape here on Earth.

As each evening progresses, though, we'll start to see brighter stars and constellations rise in the sky. Leading the way is **Taurus** (the bull) with the lovely little star cluster of the **Pleiades** (the Seven Sisters). Accompanying the bull is the charioteer – the constellation **Auriga** with its bright leading star **Capella**.

Following on are **Orion** and **Gemini**. They are traditionally the constellations of winter, and a reminder that the hardest season is still to come – but they also bring with them the promise of more wonderful skysights.

▼ The sky at 10 pm in the middle of November, with Moon-positions marked at intervals of three days. The positions of the stars are also correct for 11 pm

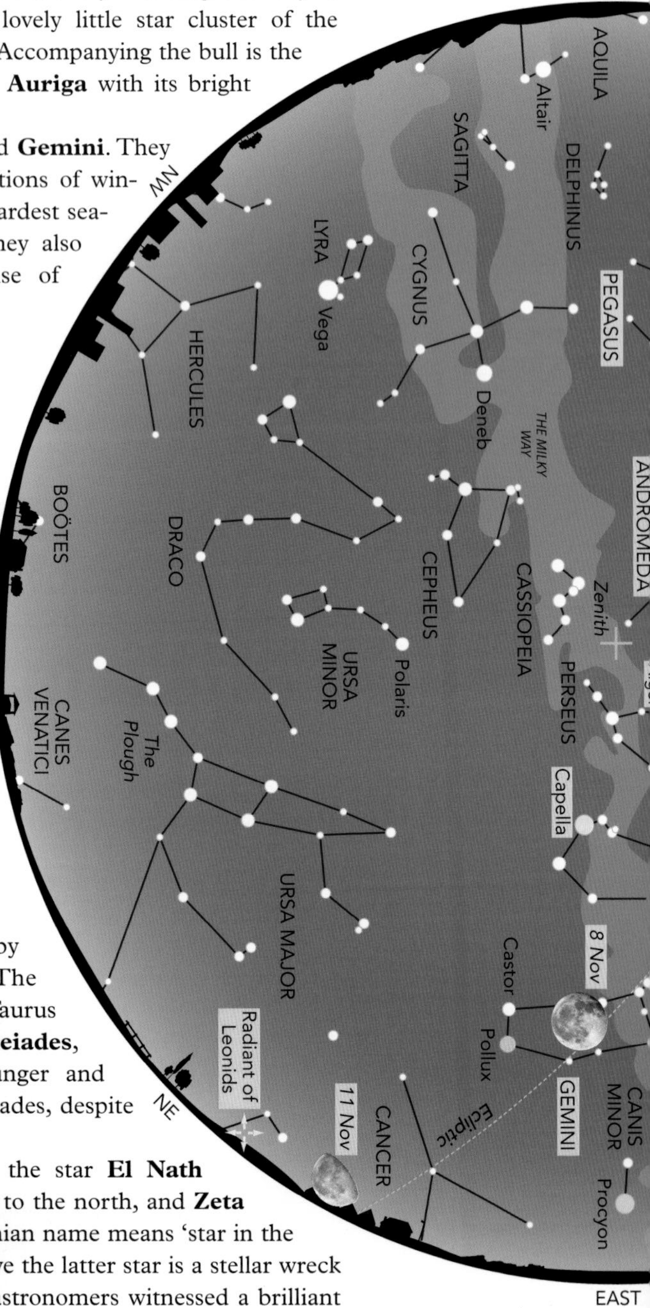

NOVEMBER'S CONSTELLATION

Taurus is very much a second cousin to brilliant Orion, but a fascinating constellation nonetheless. It is dominated by **Aldebaran**, the baleful blood-red eye of the celestial bull. Around 68 light years away, and shining with a magnitude of +0.85, Aldebaran is a red giant star, but not one as extreme as neighbouring **Betelgeuse**. It is around three times heavier than the Sun.

The bull's 'head' is formed by the **Hyades** star cluster. The other famous star cluster in Taurus is the far more glamorous **Pleiades**, the stars of which are younger and brighter than those of the Hyades, despite being further away.

Taurus has two 'horns' – the star **El Nath** (Arabic for 'the butting one') to the north, and **Zeta** (its unpronounceable Babylonian name means 'star in the bull towards the south'). Above the latter star is a stellar wreck – literally. In 1054, Chinese astronomers witnessed a brilliant

at the beginning of November, and 9 pm at the end of the month. The planets move slightly relative to the stars during the course of the month.

'new star' appear in this spot, which was visible in daytime for weeks. What the Chinese actually saw was an exploding star – a *supernova* – in its death throes. Today, we see its still-expanding remains as the **Crab Nebula** (M1). It is visible through a medium-sized telescope.

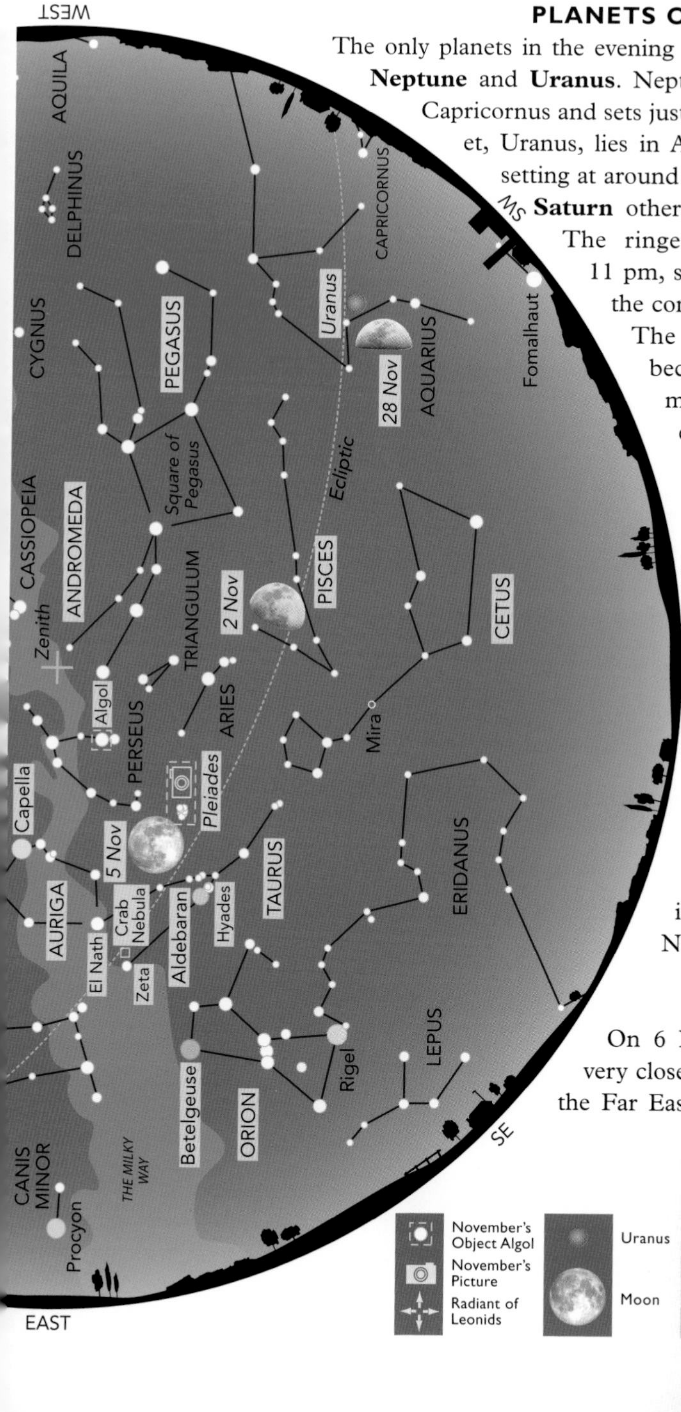

PLANETS ON VIEW

The only planets in the evening sky are the dim outer giants **Neptune** and **Uranus**. Neptune (magnitude +7.9) is in Capricornus and sets just before 10 pm. Its twin planet, Uranus, lies in Aquarius at magnitude +5.8, setting at around 0.30 am.

Saturn otherwise has the night to itself. The ringed world rises just before 11 pm, shining at magnitude +0.5 in the constellation of Leo.

The two inner planets both become visible at the end of the month. **Venus** appears in the evening sky in the southwest, shining at magnitude –3.9.

Mercury is putting on its best morning performance of the year: it is at its greatest eastern elongation on 25 November. In order to spot the elusive little planet (shining at magnitude –0.5), look low on the southeastern horizon at around 6 am during the last two weeks of November.

Mars and **Jupiter** are lost in the Sun's glare all through November.

MOON

On 6 November, the Moon rises very close to the **Pleiades** (people in the Far East will have seen the Moon

MOON		
Date	Time	Phase
5	12.58 pm	Full Moon
12	5.45 pm	Last Quarter
20	10.18 pm	New Moon
28	6.29 am	First Quarter

hide some of its stars). The night of 12/13 November sees the Last Quarter Moon passing above the planet Saturn. In the early morning of 18 November, the thin crescent Moon lies to the right of Mercury. After it passes the Sun, a crescent Moon appears to the left of Venus in the dusk twilight of 22 November. On 28 November, the Moon lies very near Uranus: as seen in binoculars, the distant planet is the fainter of the two 'stars' to the right of the First Quarter Moon.

SPECIAL EVENTS

17–19 November: This is the maximum of the **Leonid** meteor shower. A few years ago this annual shower yielded literally storms of shooting stars – many thousands an hour. The number is much reduced now, as the parent comet Tempel–Tuttle, which sheds its dust to produce the meteors, has moved away from the vicinity of Earth. While the traditional time for maximum is the night of 17 November, David Asher from Armagh Observatory predicts that there may be a second sharp peak – with perhaps 100 meteors per hour – at 4.45 am on 19 November.

NOVEMBER'S OBJECT

The star **Algol**, in the constellation Perseus, represents the head of the dreadful Gorgon Medusa. In Arabic, its name

Viewing tip
Now that the nights are drawing in and becoming darker, it's a good time to pick out faint, fuzzy objects like the Andromeda Galaxy and the Orion Nebula. But don't even think about it near the time of Full Moon, as its light will drown them out. The best time to observe 'deep sky objects' is when the Moon is near to New, or after Full Moon. Check the Moon phases timetable in this book.

means 'the Demon'. Watching Algol carefully reveals why. Every 2 days 21 hours, Algol dims in brightness for several hours – to become as faint as the star lying to its lower right (Gorgonea Tertia).

In 1783, a young British amateur astronomer, John Goodricke of York, discovered Algol's regular changes, and proposed that Algol is orbited by a large dark planet that periodically blocks off some of its light. We now know that Algol does indeed have a dim companion blocking its brilliant light, but it is a fainter star rather than a planet.

Many of the minima of Algol take place during daylight; but keep a spot check on Algol over several hours on the nights of 9/10, 11/12 and 14/15 November to catch the cosmic wink.

NOVEMBER'S PICTURE

The **Pleiades** (Seven Sisters) star cluster rises early in the evenings in November. It has fascinated people around the world from prehistoric times. The Chinese recorded the Pleiades as early as 2357 BC. The Polynesians used this cluster as a navigational aid when crossing the empty Pacific Ocean; they saw the Pleiades as the eyes of Rigi, a worm-god who tried to raise the heavens but broke apart under the strain.

NOVEMBER'S TOPIC
Mars Reconnaissance Orbiter

This month sees the start of operations for a phenomenal new Mars orbiter, which should have reached the planet in March. NASA's Mars Reconnaissance Orbiter (MRO) is bristling with sophisticated instruments, its technology the fastest ever used on a space probe. The huge two-tonne craft will send back images and information at a rate quicker than an internet link, with the result that researchers will be swamped with new data.

MRO is looking in particular for water on Mars. It has an Italian radar experiment on board which will penetrate the Red Planet's surface, looking for water actually under the soil (where most researchers believe it lies). MRO will also search for evidence of ancient oceans and hot springs – as well as taking a long, hard look at the Martian atmosphere. Its cameras will be capable of imaging objects as small as a coffee table – so look forward to some amazing discoveries!

◀ Another beautiful composite picture by Robin Scagell. Against a twilight sky foreground, he superimposed an image of the Pleiades taken with a 135 mm telephoto lens on Kodachrome 64. The exposure was 10 seconds.

Matching the bright lights of the Christmas shops, the sky is putting on its annual display of luminous stars. The great hunter **Orion** alone contains 10% of the brightest 70 stars in the sky. He's accompanied by the most brilliant star of all, **Sirius**, the Dog Star in the constellation of **Canis Major** (the great dog).

Other 1st-magnitude stars are to be found surrounding Orion, in a great arc that includes **Procyon** (in **Canis Minor**, the little dog), **Castor** and **Pollux** (**Gemini**, the twins), **Capella** (**Auriga**, the charioteer) and **Aldebaran** (**Taurus**, the bull).

The seasonal sky-show will be enhanced by the Seven Sisters (**Pleiades**) doing a disappearing act and by a shower of shooting stars.

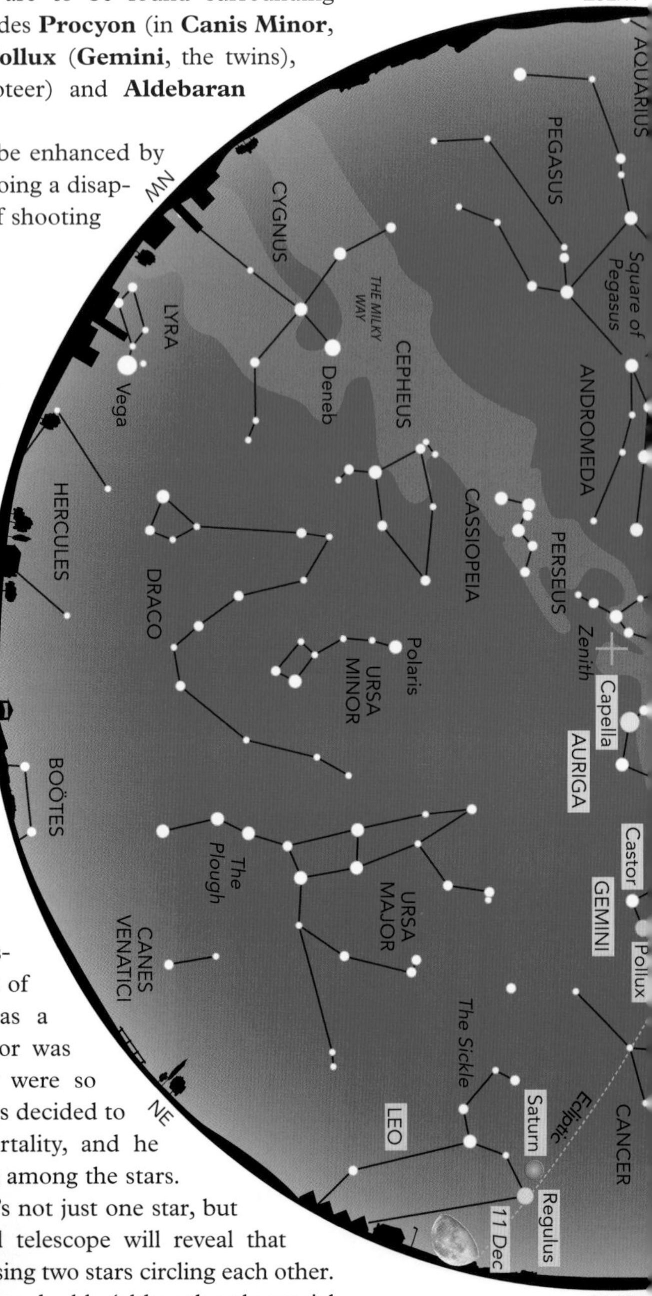

▼ *The sky at 10 pm in the middle of December, with Moon-positions marked at intervals of three days. The positions of the stars are also correct for 11 pm*

DECEMBER'S CONSTELLATION

You can't ignore **Gemini** in December. High in the southeast, the constellation is dominated by the stars **Castor** and **Pollux**. They are of similar brightness and represent the heads of a pair of twins, with their stellar bodies running in parallel lines of stars towards the west.

Legend has it that Castor and Pollux were twins, conceived on the same night by the princess Leda. On the night that she married the King of Sparta, wicked old Zeus (Jupiter) invaded the marital suite, disguised as a swan. Pollux, born of the liaison with Zeus, was as a result immortal – while Castor was merely human. But the pair were so devoted to each other that Zeus decided to grant Castor honorary immortality, and he placed both Castor and Pollux among the stars.

Castor is an amazing star. It's not just one star, but a family of six. Even a small telescope will reveal that Castor is a double star, comprising two stars circling each other. Both of these stars are themselves double (although only special

t the beginning of December, nd 9 pm at the end of the nonth. The planets move slightly elative to the stars during the ourse of the month.

equipment can detect this). In addition there's an outlying star, visible through a telescope, which also turns out to be double.

PLANETS ON VIEW

After swinging round behind the Sun, **Venus** is now reappearing in the evening sky. Look low down in the southwest, towards the end of December, to spot the Evening Star shining at magnitude –3.9.

Saturn is now dominating the night skies, rising at around 9 pm in the constellation **Leo** (the lion). At magnitude +0.4, the planet is twice as bright as Leo's most prominent star, **Regulus**. Through binoculars the pair make an interesting contrast, the yellow steady glow of Saturn contrasting with the blue-white twinkling star.

Neptune (magnitude +7.9), in Capricornus, sets at around 8.30 pm; while **Uranus** (magnitude +5.8), in Aquarius, follows it below the horizon just before 11 pm.

There's more action in the dawn skies. At the beginning of December, you may catch **Mercury** (magnitude –0.5) low in the southeast before 7 am. As it sinks into the twilight glow, Mercury meets the much brighter **Jupiter** (magnitude –1.7) coming the other way. **Mars** is also on the scene, though much fainter, at magnitude +1.5. The three planets make a lovely little triangle on the mornings of 9–12 December, though unfortunately very low in the bright dawn sky. As the month goes by, Jupiter rises ever higher in the morning sky.

MOON		
Date	Time	Phase
5	12.58 pm	Full Moon
12	2.32 pm	Last Quarter
20	2.01 pm	New Moon
27	2.48 pm	First Quarter

MOON

On 3/4 December, the Moon occults the **Pleiades**. On the evenings of 9 and 10 December, the Moon lies near Saturn. The thin crescent Moon lies to the right of Jupiter in the dawn sky on 18 December. On the night of 25 December, the Moon makes its closest pass of the year to Uranus – at around 9 pm, train your binoculars on what looks like a star less than one moonwidth above the Moon itself.

▶ *The unique Crab Nebula, as captured by Michael Stecker in California. He used a Celestron C-14 telescope at f/6 and Konica 1600 film.*

SPECIAL EVENTS

3/4 December: There's an occultation of the **Pleiades**. The small hours of 4 December see a rare event as the almost-Full Moon moves right across in front of the Pleiades star cluster: between 2.50 and 4.50 am, six out of the Seven Sisters will be occulted. The Moon itself will be so bright that you will need binoculars or a small telescope to watch the action.

5/6 December: The Full Moon tonight is the highest in 18 years (altitude 66½ degrees).

13 December: It is the maximum of the **Geminid** meteor shower, which lasts from 7 to 16 December. The meteors are debris shed from an asteroid called Phaethon, and are therefore quite substantial – and hence bright. This is a good year for observing them, as moonlight won't interfere until well after midnight.

22 December 0.22 am: Today is the Winter Solstice. As a result of the tilt of Earth's axis, the Sun reaches its lowest point in the heavens as seen from the northern hemisphere: we get the shortest days, and the longest nights.

DECEMBER'S OBJECT

Look at Orion's Belt, and – on a clear night – you'll detect a small fuzzy patch below the line of stars. Through binoculars, or a small telescope, the patch looks like a small cloud in space. The **Orion Nebula** is indeed a cloud, but at 30 light years across, it's hardly small. Only its distance – 1500 light years – diminishes it. Yet it is one of the the nearest regions of star formation to Earth, containing at least 150 fledgling stars (known as *protostars*), which have condensed out of the gas. This 'star factory' is lit by fierce radiation from a small cluster of newly born stars called 'the Trapezium', which are beautiful to look at through a small telescope. The Orion Nebula is just part of a huge gas complex in the Orion region which may have enough material to make 500,000 stars in the future.

DECEMBER'S PICTURE

The **Crab Nebula** (M1), in Taurus, is the remains of a star that exploded in 1054. It is just visible through a small telescope, but only a big instrument can reveal its structure, which

Viewing tip

Venus is dazzling this month, looking like a Christmas tree light hanging in the sky. If you have a small telescope, though, don't wait until the sky gets totally dark – as Venus will appear so bright that it will be difficult to make out anything. It is best to wait until just after the Sun has set (not before, or you risk catching the Sun in your field of view), when you'll see the crescent Venus appearing fainter against a pale blue sky.

astronomer Lord Rosse, in 1848, compared to the pincers of a crab. This supernova remnant is 15 light years across, and still expanding.

DECEMBER'S TOPIC
The expanding Universe

Christmas is coming up, with its associations with birth and beginnings. But how did our Universe begin? Luckily, there is some pretty firm evidence on which to base an answer. Firstly, the Universe is expanding – galaxies are moving apart from each other. 'Rewinding the tape' reveals that the expansion dates back to a time 13.7 billion years ago – a measurement that has only been tied down in recent years. Secondly, the Universe is not entirely cold: it's bathed in a radiation field with a temperature of 2.7 degrees above Absolute Zero. Both these clues point to the origin of the Universe in a blisteringly hot 'Big Bang', which caused space to expand. The 'microwave background' of 2.7 degrees is the remnant of this birth in fire, cooled down by the relentless expansion to a mere shadow of its former self. Current observations show that the Universe is not just expanding, but accelerating – which means that it's destined to die by simply fading away rather than re-collapsing in a cataclysmic 'Big Crunch'.

There's always something to see in our Solar System, from planets to meteors or the Moon. These objects are very close to us – in astronomical terms – so their positions, shapes and sizes appear to change constantly. It is important to know when, where and how to look if you are to enjoy exploring Earth's neighbourhood. Here we give the best dates in 2006 for observing the planets and meteors (weather permitting!), and explain some of the concepts that will help you to get the most out of your observing.

THE INFERIOR PLANETS

A planet with an orbit that lies closer to the Sun than the orbit of Earth is known as *inferior*. Mercury and Venus are the inferior planets. They show a full range of phases (like the Moon) from the thinnest crescents to full, depending on their position in relation to the Earth and the Sun. The diagram shows the various positions of the inferior planets. They are invisible at *conjunction* and best viewed at their eastern or western *elongations*.

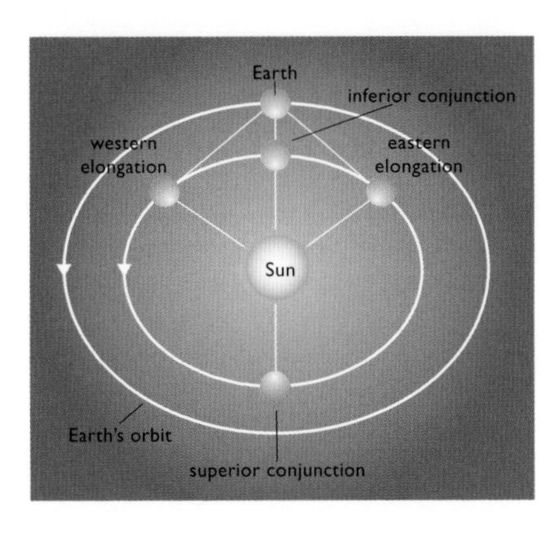

◀ *At eastern or western elongation, an inferior planet is at its maximum angular distance from the Sun. Conjunction occurs at two stages in the planet's orbit. Under certain circumstances an inferior planet can transit across the Sun's disc at inferior conjunction.*

Mercury is close to the Sun and is visible only for a period of roughly a week, six to eight times a year. From the northern hemisphere, it is best spotted around sunset in the spring during an eastern elongation from the Sun, and around dawn in the autumn during a western elongation. Sweep the approximate area of the sky with binoculars to locate the planet soon after sunset.

In late February 2006, Mercury is at its greatest elongation (east) of the Sun and is visible during dusk after sunset. In late June, it is again visible in the early evening. In late November, it is at its greatest elongation (west) and is visible in the dawn before sunrise.

Maximum elongations of Mercury in 2006

Date	Time (UT)	Separation
24 Feb	05:03	18° 07' 32" east
8 Apr	18:38	27° 45' 49" west
20 June	20:10	24° 56' 16" east
7 Aug	00:32	19° 11' 11" west
17 Oct	04:07	24° 49' 14" east
25 Nov	12:56	19° 54' 16" west

Venus lies farther from the Sun than Mercury. Its phases can be seen in good binoculars or a small telescope, while a larger telescope is needed to see features in the upper atmosphere. Venus is at its brightest about 36 days either side of inferior conjunction.

From mid-January until September 2006, Venus is a prominent morning object, and is at its greatest elongation (west) in late March. From November it is visible in the early evening.

Maximum elongation of Venus in 2006

Date	Time (UT)	Separation
25 Mar	06:44	46° 31' 49" west

► *Superior planets are invisible at conjunction. At quadrature the planet is at right angles to the Sun as viewed from Earth. Opposition is the best time to observe a superior planet.*

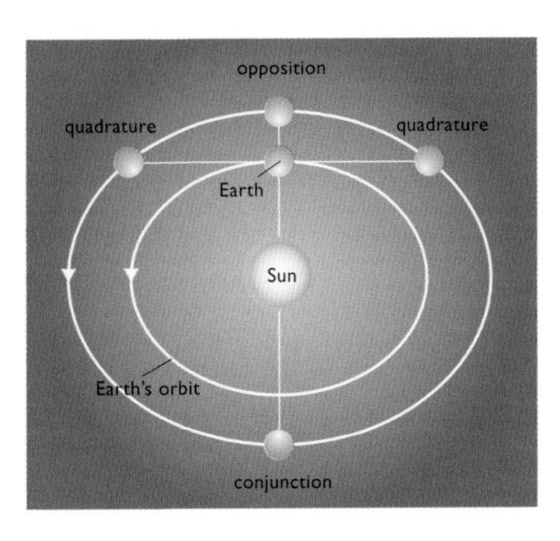

THE SUPERIOR PLANETS

The superior planets are those with orbits that lie beyond that of the Earth. They are Mars, Jupiter, Saturn, Uranus, Neptune and Pluto. The best time to observe a superior planet is when the Earth lies between it and the Sun. At this point in the planet's orbit, it is said to be at *opposition*.

Progress of Mars through the constellations	
1 Jan–7 Feb	Aries
8 Feb–14 Apr	Taurus
15 Apr–31 May	Gemini
1 June–2 July	Cancer
3 July–29 Aug	Leo
30 Aug–4 Nov	Virgo
5 Nov–7 Dec	Libra
8 Dec–17 Dec	Scorpius
18 Dec–31 Dec	Ophiuchus

Mars is a disappointing sight through a small telescope. Larger amateur telescopes, however, reveal the dark markings on the surface, changes in the polar ice caps, as well as Mars' phases and the progress of dust storms. Because Mars rotates in approximately 24 hours and 37 minutes, the features change during the night.

● *Mars is best placed for observation at the beginning of the year, becoming gradually fainter from January to June. It is not visible from July to November, being at conjunction with the Sun on 23 October 2006.*

Magnitudes

Astronomers measure the brightness of stars, planets and other celestial objects using a scale of *magnitudes*. Somewhat confusingly, fainter objects have higher magnitudes, while brighter objects have lower magnitudes; the most brilliant stars have negative magnitudes! Naked-eye stars range from magnitude –1.5 for the brightest star, Sirius, to +6.5 for the faintest stars you can see on a really dark night. As a guide, here are the magnitudes of selected objects:

Object	Magnitude
Sun	–26.7
Full Moon	–12.5
Venus (at its brightest)	–4.6
Sirius	–1.5
Betelgeuse	+0.4
Polaris (Pole Star)	+2.0
Faintest star visible to the naked eye	+6.5
Pluto	+14
Faintest star visible to the Hubble Space Telescope	+31

Jupiter is a very rewarding target. Even a small telescope will reveal some of the bands in its atmosphere and the Great Red Spot. Larger instruments will reveal many of the smaller features in the clouds, such as white spots. The 10-hour rotation period of Jupiter's upper atmosphere makes it an ideal candidate for short-exposure CCD or video imaging. Drawing Jupiter can be challenging because of the rapid rotation.

The four largest satellites of Jupiter – Callisto, Europa, Ganymede and Io – can be seen in most binoculars. Their orbits can be plotted using even moderate-sized telescopes and their transits and occultations can be observed.

● *Jupiter lies in the constellation of Libra through most of 2006. It is best placed for observation in the early summer, particularly around the time it reaches opposition on 4 May.*

Saturn is perhaps best known for its ring system, which can be seen in small telescopes. Larger telescopes will reveal the structure of the main rings, the banding of the planet's atmosphere and changes such as the appearance of white spots.

The angle at which the rings are seen changes over a 30-year period. Recording this cycle makes an interesting long-term project. Some of Saturn's brightest satellites – Dione, Iapetus,

Rhea, Titan and Enceladus – are visible in larger telescopes. Like the Galilean satellites of Jupiter, their orbits can be plotted and their transits observed.

- *Saturn starts 2006 in the constellation of Cancer and moves into Leo in September. It is at opposition on 27 January and is well placed for observation from January to April and November to December.*

Uranus is just visible with the naked eye, and is better seen with good binoculars or a small telescope. With a powerful telescope, it appears as a small blue-green disc. The brighter satellites can be picked up in CCD images.

- *Throughout 2006 Uranus lies in the constellation of Aquarius, though for much of the early part of the year it is drowned out in the glare from the Sun. Visibility improves during the summer, and it reaches opposition on 5 September.*

Neptune is visible in good binoculars or a small telescope, and is a pale blue-green. Its largest satellite, Triton, may be seen in large amateur instruments.

- *Neptune spends 2006 in the constellation of Capricornus. It is not visible early in the year, and is best viewed in the summer. By July, Neptune is above the horizon all night, and it reaches opposition on 10 August.*

Pluto is very faint, magnitude 14, and difficult to locate even in the largest amateur instruments. Some astronomers do not even regard it as a planet.

SOLAR AND LUNAR ECLIPSES

Solar eclipses, particularly total solar eclipses, are among the most exciting sky sights you can see. A solar eclipse occurs when the Moon's shadow falls on Earth.

For a solar eclipse to occur, the Moon must be perfectly in line between the Sun and Earth. This happens at least twice a year, and sometimes as many as five times. However, you do have to be in the right location on Earth to see the eclipse.

- *There are two solar eclipses in 2006, on 29 March and 22 September. The former (a total eclipse) will be seen from northern South America, North Africa and central Asia; it will appear as a partial eclipse throughout Europe, western Asia and most of Africa. The latter (an annular eclipse) will be visible from the South Atlantic.*

A **lunar eclipse** occurs when the Moon passes through Earth's shadow. As with solar eclipses, it may be total or partial, depending on whether the Moon passes completely or partially through the dark central part of Earth's shadow.

- *There is a partial lunar eclipse on 7 September 2006. It will be visible from Asia, Europe and eastern Africa, though not in North America.*

Astronomical distances

For objects in the Solar System, like the planets, we can give their distances from the Earth in kilometres. But the distances are just too huge once we reach out to the stars. Even the nearest star (Proxima Centauri) lies 25 million million km away. So astronomers use a larger unit, the *light year*. This is the distance that light travels in one year, and it equals 9.46 million million km.

Here are the distances to some familiar astronomical objects, in light years:

Proxima Centauri	4.2
Betelgeuse	427
Centre of the Milky Way	24,000
Andromeda Galaxy	2.9 million
Most distant galaxies seen by the Hubble Space Telescope	13 billion

Angular separations

Astronomers measure the distance between objects, as we see them in the sky, by the angle between the objects, in degrees (symbol °). From the horizon to the point above your head is 90 degrees. All around the horizon is 360 degrees.

You can use your hand, held at arm's length, as a rough guide to angular distances, as follows:

Width of index finger	1°
Width of clenched hand	10°
Thumb to little finger on outspread hand	20°

For smaller distances, astronomers divide the degree into 60 arcminutes (symbol ′), and the arcminute into 60 arcseconds (symbol ″).

▶ *Where the dark central part (the umbra) of the Moon's shadow reaches the Earth, a total eclipse is seen. People located within the penumbra see a partial eclipse. If the umbral shadow does not reach Earth, an annular eclipse is seen. This type of eclipse occurs when the Moon is at a distant point in its orbit and is not quite large enough to cover the whole of the Sun's disc.*

Annular solar eclipse
Partial eclipse
Annular eclipse
Umbra
Sun
Moon
View from Earth
Penumbra
Total solar eclipse
Total eclipse

Dates of maximum for selected meteor showers	
Meteor shower	date of maximum
Quadrantids	3 January
Lyrids	22 April
Eta Aquarids	5 May
Perseids	12/13 August
Orionids	21 October
Leonids	17–19 November
Geminids	13 December

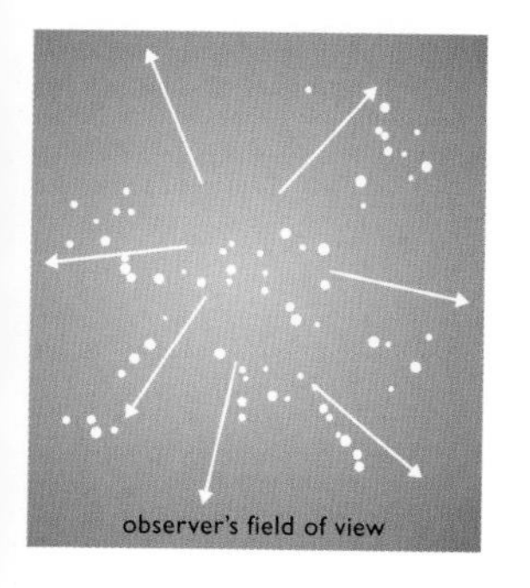

▲ *Meteors from a common source, occurring during a shower, enter the atmosphere along parallel trajectories. As a result of perspective, however, they appear to diverge from a single point in the sky.*

METEOR SHOWERS

Shooting stars – or *meteors* – are tiny particles of interplanetary dust, known as *meteoroids*, burning up in the Earth's atmosphere. At certain times of year, the Earth passes through a stream of these meteoroids (usually debris left behind by a comet) and a *meteor shower* is seen. The point in the sky from which the meteors appear to emanate is known as the *radiant*. Most showers are known by the constellation in which the radiant is situated.

When watching meteors for a coordinated meteor programme, observers generally note the time, seeing conditions, cloud cover, their own location, the time and brightness of each meteor and whether it was from the main meteor stream or not. It's also worth noting any details of persistent afterglows (trains) and fireballs, and making counts of how many meteors appear in a given period.

COMETS

Comets are small bodies in orbit about the Sun. Consisting of frozen gases and dust, they are often known as 'dirty snowballs'. When their orbits bring them close to the Sun the ices evaporate and dramatic tails of gas and dust can sometimes be seen.

A number of comets move round the Sun in fairly small, elliptical orbits in periods of a few years; others have much longer periods. Most really brilliant comets have orbital periods of several thousands or even millions of years. The exception is Comet Halley, a bright comet with a period of about 76 years. It was last seen in 1986.

Binoculars and wide-field telescopes provide the best views of comet tails. Larger telescopes with a high magnification are necessary to observe fine detail in the gaseous head (coma). Most comets are discovered with professional instruments, but a few are still found by experienced amateur astronomers.

Deep sky objects are 'fuzzy patches' that lie outside the Solar System. They include star clusters, nebulae and galaxies. To observe the majority of deep sky objects you will need binoculars or a telescope, but there are also some beautiful naked eye objects, notably the Pleiades and the Orion Nebula.

The faintest object that an instrument can see is its *limiting magnitude*. The table gives a rough guide, for good seeing conditions, for a variety of small- to medium-sized telescopes.

We have provided a selection of recommended deep sky targets, together with their magnitudes. Some are described in more detail in our 'Object of the month' features. Look on the appropriate month's map to find which constellations are on view, and then choose your objects using the list below. We have provided celestial coordinates, for readers with detailed star maps. The suggested times of year for viewing are when the constellation is highest in the sky in the late evening.

Limiting magnitude for small to medium telescopes

aperture (mm)	limiting magnitude
50	+11.2
60	+11.6
70	+11.9
80	+12.2
100	+12.7
125	+13.2
150	+13.6

RECOMMENDED DEEP SKY OBJECTS

Andromeda – autumn and early winter	
M31 (NGC 224) Andromeda Galaxy	3rd magnitude spiral galaxy, *RA 00h 42.7m Dec +41° 16'*
M32 (NGC 221)	8th magnitude elliptical galaxy, a companion to M31. *RA 00h 42.7m Dec +40° 52'*
M110 (NGC 205)	8th magnitude elliptical galaxy *RA 00h 40.4m Dec +41° 41'*
NGC 7662 Blue Snowball	8th magnitude planetary nebula *RA 23h 25.9m Dec +42° 33'*
Aquarius – late autumn and early winter	
M2 (NGC 7089)	6th magnitude globular cluster *RA 21h 33.5m Dec –00° 49'*
M72 (NGC 6981)	9th magnitude globular cluster *RA 20h 53.5m Dec –12° 32'*
NGC 7293 Helix Nebula	7th magnitude planetary nebula *RA 22h 29.6m Dec –20° 48'*
NGC 7009 Saturn Nebula	8th magnitude planetary nebula; *RA 21h 04.2m Dec –11° 22'*
Aries – early winter	
NGC 772	10th magnitude spiral galaxy *RA 01h 59.3m Dec +19° 01'*
Auriga – winter	
M36 (NGC 1960)	6th magnitude open cluster *RA 05h 36.1m Dec +34° 08'*
M37 (NGC 2099)	6th magnitude open cluster *RA 05h 52.4m Dec +32° 33'*
M38 (NGC 1912)	6th magnitude open cluster *RA 05h 28.7m Dec +35° 50'*
Cancer – late winter to early spring	
M44 (NGC 2632) Praesepe or Beehive	3rd magnitude open cluster *RA 08h 40.1m Dec +19° 59'*
M67 (NGC 2682)	7th magnitude open cluster *RA 08h 50.4m Dec +11° 49'*
Canes Venatici – visible all year	
M3 (NGC 5272)	6th magnitude globular cluster *RA 13h 42.2m Dec +28° 23'*
M51 (NGC 5194/5) Whirlpool Galaxy	8th magnitude spiral galaxy *RA 13h 29.9m Dec +47° 12'*
M63 (NGC 5055)	9th magnitude spiral galaxy *RA 13h 15.8m Dec +42° 02'*
M94 (NGC 4736)	8th magnitude spiral galaxy *RA 12h 50.9m Dec +41° 07'*
M106 (NGC4258)	8th magnitude spiral galaxy *RA 12h 19.0m Dec +47° 18'*
Canis Major – late winter	
M41 (NGC 2287)	4th magnitude open cluster *RA 06h 47.0m Dec –20° 44'*
Capricornus – late summer and early autumn	
M30 (NGC 7099)	7th magnitude globular cluster *RA 21h 40.4m Dec –23° 11'*
Cassiopeia – visible all year	
M52 (NGC 7654)	6th magnitude open cluster *RA 23h 24.2m Dec +61° 35'*
M103 (NGC 581)	7th magnitude open cluster *RA 01h 33.2m Dec +60° 42'*
NGC 225	7th magnitude open cluster *RA 00h 43.4m Dec +61 47'*
NGC 457	6th magnitude open cluster *RA 01h 19.1m Dec +58° 20'*
NGC 663	Good binocular open cluster *RA 01h 46.0m Dec +61° 15'*
Cepheus – visible all year	
Delta Cephei	Variable star, varying between 3.5 and 4.4 with a period of 5.37 days. It has a magnitude 6.3 companion and they make an attractive pair for small telescopes or binoculars.
Cetus – late autumn	
Mira (omicron Ceti)	Irregular variable star with a period of roughly 330 days and a range between 2.0 and 10.1.
M77 (NGC 1068)	9th magnitude spiral galaxy *RA 02h 42.7m Dec –00° 01'*

Coma Berenices – spring

M53 (NGC 5024)	8th magnitude globular cluster *RA 13h 12.9m Dec +18° 10'*
M64 (NGC 4286) Black Eye Galaxy	8th magnitude spiral galaxy with a prominent dust lane that is visible in larger telescopes. *RA 12h 56.7m Dec +21° 41'*
M85 (NGC 4382)	9th magnitude elliptical galaxy *RA 12h 25.4m Dec +18° 11'*
M88 (NGC 4501)	10th magnitude spiral galaxy *RA 12h 32.0m Dec.+14° 25'*
M91 (NGC 4548)	10th magnitude spiral galaxy *RA 12h 35.4m Dec +14° 30'*
M98 (NGC 4192)	10th magnitude spiral galaxy *RA 12h 13.8m Dec +14° 54'*
M99 (NGC 4254)	10th magnitude spiral galaxy *RA 12h 18.8m Dec +14° 25'*
M100 (NGC 4321)	9th magnitude spiral galaxy *RA 12h 22.9m Dec +15° 49'*
NGC 4565	10th magnitude spiral galaxy *RA 12h 36.3m Dec +25° 59'*

Cygnus – late summer and autumn

Cygnus Rift	Dark cloud just south of Deneb that appears to split the Milky Way in two.
NGC 7000 North America Nebula	A bright nebula against the background of the Milky Way, visible with binoculars under dark skies. *RA 20h 58.8m Dec +44° 20'*
NGC 6992 Veil Nebula (part)	Supernova remnant, visible with binoculars under dark skies. *RA 20h 56.8m Dec +31 28'*
M29 (NGC 6913)	7th magnitude open cluster *RA 20h 23.9m Dec +36° 32'*
M39 (NGC 7092)	Large 5th magnitude open cluster *RA 21h 32.2m Dec +48° 26'*
NGC 6826 Blinking Planetary	9th magnitude planetary nebula *RA 19 44.8m Dec +50° 31'*

Delphinus – late summer

NGC 6934	9th magnitude globular cluster *RA 20h 34.2m Dec +07° 24'*

Draco – midsummer

NGC 6543	9th magnitude planetary nebula *RA 17h 58.6m Dec +66° 38'*

Gemini – winter

M35 (NGC 2168)	5th magnitude open cluster *RA 06h 08.9m Dec +24° 20'*
NGC 2392 Eskimo Nebula	8–10th magnitude planetary nebula *RA 07h 29.2m Dec +20° 55'*

Hercules – early summer

M13 (NGC 6205)	6th magnitude globular cluster *RA 16h 41.7m Dec +36° 28'*
M92 (NGC 6341)	6th magnitude globular cluster *RA 17h 17.1m Dec +43° 08'*
NGC 6210	9th magnitude planetary nebula *RA 16h 44.5m Dec +23 49'*

Hydra – early spring

M48 (NGC 2548)	6th magnitude open cluster *RA 08h 13.8m Dec –05° 48'*
M68 (NGC 4590)	8th magnitude globular cluster *RA 12h 39.5m Dec –26° 45'*
M83 (NGC 5236)	8th magnitude spiral galaxy *RA 13h 37.0m Dec –29° 52'*
NGC 3242 Ghost of Jupiter	9th magnitude planetary nebula *RA 10h 24.8m Dec –18°38'*

Leo – spring

M65 (NGC 3623)	9th magnitude spiral galaxy *RA 11h 18.9m Dec +13° 05'*
M66 (NGC 3627)	9th magnitude spiral galaxy *RA 11h 20.2m Dec +12° 59'*
M95 (NGC 3351)	10th magnitude spiral galaxy *RA 10h 44.0m Dec +11° 42'*
M96 (NGC 3368)	9th magnitude spiral galaxy *RA 10h 46.8m Dec +11° 49'*
M105 (NGC 3379)	9th magnitude elliptical galaxy *RA 10h 47.8m Dec +12° 35'*

Lepus – winter

M79 (NGC 1904)	8th magnitude globular cluster *RA 05h 24.5m Dec –24° 33'*

Lyra – spring

M56 (NGC 6779)	8th magnitude globular cluster *RA 19h 16.6m Dec +30° 11'*
M57 (NGC 6720) Ring Nebula	9th magnitude planetary nebula *RA 18h 53.6m Dec +33° 02'*

Monoceros – winter

M50 (NGC 2323)	6th magnitude open cluster *RA 07h 03.2m Dec –08° 20'*
NGC 2244	Open cluster surrounded by the faint Rosette Nebula, NGC 2237. Visible in binoculars. *RA 06h 32.4m Dec +04° 52'*

Ophiuchus – summer

M9 (NGC 6333)	8th magnitude globular cluster *RA 17h 19.2m Dec –18° 31'*
M10 (NGC 6254)	7th magnitude globular cluster *RA 16h 57.1m Dec –04° 06'*
M12 (NCG 6218)	7th magnitude globular cluster *RA 16h 47.2m Dec –01° 57'*
M14 (NGC 6402)	8th magnitude globular cluster *RA 17h 37.6m Dec –03° 15'*
M19 (NGC 6273)	7th magnitude globular cluster *RA 17h 02.6m Dec –26° 16'*
M62 (NGC 6266)	7th magnitude globular cluster *RA 17h 01.2m Dec –30° 07'*
M107 (NGC 6171)	8th magnitude globular cluster *RA 16h 32.5m Dec –13° 03'*

Orion – winter

M42 (NGC 1976) Orion Nebula	4th magnitude nebula *RA 05h 35.4m Dec –05° 27'*
M43 (NGC 1982)	5th magnitude nebula *RA 05h 35.6m Dec –05° 16'*
M78 (NGC 2068)	8th magnitude nebula *RA 05h 46.7m Dec +00° 03'*

Pegasus – autumn

M15 (NGC 7078)	6th magnitude globular cluster *RA 21h 30.0m Dec +12° 10'*

Perseus – autumn to winter

M34 (NGC 1039)	5th magnitude open cluster *RA 02h 42.0m Dec +42° 47'*
M76 (NGC 650/1) Little Dumbbell	11th magnitude planetary nebula *RA 01h 42.4m Dec +51° 34'*

NGC 869/884 Double Cluster	Pair of open star clusters *RA 02h 19.0m Dec +57° 09'* *RA 02h 22.4m Dec +57° 07'*
Pisces – autumn	
M74 (NGC 628)	9th magnitude spiral galaxy *RA 01h 36.7m Dec +15° 47'*
Puppis – late winter	
M46 (NGC 2437)	6th magnitude open cluster *RA 07h 41.8m Dec –14° 49'*
M47 (NGC 2422)	4th magnitude open cluster *RA 07h 36.6m Dec –14° 30'*
M93 (NGC 2447)	6th magnitude open cluster *RA 07h 44.6m Dec –23° 52'*
Sagitta – late summer	
M71 (NGC 6838)	8th magnitude globular cluster *RA 19h 53.8m Dec +18° 47'*
Sagittarius – summer	
M8 (NGC 6523) Lagoon Nebula	6th magnitude nebula *RA 18h 03.8m Dec –24° 23'*
M17 (NGC 6618) Omega Nebula	6th magnitude nebula *RA 18h 20.8m Dec –16° 11'*
M18 (NGC 6613)	7th magnitude open cluster *RA 18h 19.9m Dec –17 08'*
M20 (NGC 6514) Trifid Nebula	9th magnitude nebula *RA 18h 02.3m Dec –23° 02'*
M21 (NGC 6531)	6th magnitude open cluster *RA 18h 04.6m Dec –22° 30'*
M22 (NGC 6656)	5th magnitude globular cluster *RA 18h 36.4m Dec –23° 54'*
M23 (NGC 6494)	5th magnitude open cluster *RA 17h 56.8m Dec –19° 01'*
M24 (NGC 6603)	5th magnitude open cluster *RA 18h 16.9m Dec –18° 29'*
M25 (IC 4725)	5th magnitude open cluster *RA 18h 31.6m Dec –19° 15'*
M28 (NGC 6626)	7th magnitude globular cluster *RA 18h 24.5m Dec –24° 52'*
M54 (NGC 6715)	8th magnitude globular cluster *RA 18h 55.1m Dec –30° 29'*
M55 (NGC 6809)	7th magnitude globular cluster *RA 19h 40.0m Dec –30° 58'*
M69 (NGC 6637)	8th magnitude globular cluster *RA 18h 31.4m Dec –32° 21'*
M70 (NGC 6681)	8th magnitude globular cluster *RA 18h 43.2m Dec –32° 18'*
M75 (NGC 6864)	9th magnitude globular cluster *RA 20h 06.1m Dec –21° 55'*
Scorpius (northern part) – midsummer	
M4 (NGC 6121)	6th magnitude globular cluster *RA 16h 23.6m Dec –26° 32'*
M7 (NGC 6475)	3rd magnitude open cluster *RA 17h 53.9m Dec –34° 49'*
M80 (NGC 6093)	7th magnitude globular cluster *RA 16h 17.0m Dec –22° 59'*
Scutum – mid- to late summer	
M11 (NGC 6705) Wild Duck Cluster	6th magnitude open cluster *RA 18h 51.1m Dec –06° 16'*

M26 (NGC 6694)	8th magnitude open cluster *RA 18h 45.2m Dec –09° 24'*
Serpens – summer	
M5 (NGC 5904)	6th magnitude globular cluster *RA 15h 18.6m Dec +02° 05'*
M16 (NGC 6611)	6th magnitude open cluster, surrounded by the Eagle Nebula. *RA 18h 18.8m Dec –13° 47'*
Taurus – winter	
M1 (NGC 1952) Crab Nebula	8th magnitude supernova remnant *RA 05h 34.5m Dec +22° 00'*
M45 Pleiades	1st magnitude open cluster, an excellent binocular object. *RA 03h 47.0m Dec +24° 07'*
Triangulum – autumn	
M33 (NGC 598)	6th magnitude spiral galaxy *RA 01h 33.9m Dec +30° 39'*
Ursa Major – all year	
M81 (NGC 3031)	7th magnitude spiral galaxy *RA 09h 55.6m Dec +69° 04'*
M82 (NGC 3034)	8th magnitude starburst galaxy *RA 09h 55.8m Dec +69° 41'*
M97 (NGC 3587) Owl Nebula	12th magnitude planetary nebula *RA 11h 14.8m Dec +55° 01'*
M101 (NGC 5457)	8th magnitude spiral galaxy *RA 14h 03.2m Dec +54° 21'*
M108 (NGC 3556)	10th magnitude spiral galaxy *RA 11h 11.5m Dec +55° 40'*
M109 (NGC 3992)	10th magnitude spiral galaxy *RA 11h 57.6m Dec +53° 23'*
Virgo – spring	
M49 (NGC 4472)	8th magnitude elliptical galaxy *RA 12h 29.8m Dec +08° 00'*
M58 (NGC 4579)	10th magnitude spiral galaxy *RA 12h 37.7m Dec +11° 49'*
M59 (NGC 4621)	10th magnitude elliptical galaxy *RA 12h 42.0m Dec +11° 39'*
M60 (NGC 4649)	9th magnitude elliptical galaxy *RA 12h 43.7m Dec +11° 33'*
M61 (NGC 4303)	10 magnitude spiral galaxy *RA 12h 21.9m Dec +04° 28'*
M84 (NGC 4374)	9th magnitude elliptical galaxy *RA 12h 25.1m Dec +12° 53'*
M86 (NGC 4406)	9th magnitude elliptical galaxy *RA 12h 26.2m Dec +12° 57'*
M87 (NGC 4486)	9th magnitude elliptical galaxy *RA 12h 30.8m Dec +12° 24'*
M89 (NGC 4552)	10th magnitude elliptical galaxy *RA 12h 35.7m Dec +12° 33'*
M90 (NGC 4569)	9th magnitude spiral galaxy *RA 12h 36.8m Dec +13° 10'*
M104 (NGC 4594) Sombrero Galaxy	Almost edge on 8th magnitude spiral galaxy. *RA 12h 40.0m Dec –11° 37'*
Vulpecula – late summer and autumn	
M27 (NGC 6853) Dumbbell Nebula	8th magnitude planetary nebula *RA 19h 59.6m Dec +22° 43'*

EQUIPMENT REVIEW

The telescope market is becoming big business. It's not in the same league as sports equipment, where billions are spent worldwide, but it is certainly no longer an industry run out of garages and small workshops. The major players these days are Meade in the USA and Synta in China, who in 2005 took over the other big name, Celestron. You won't see the Synta name on any telescopes, though – for as well as Celestron, they make Sky-Watcher, which is their worldwide brand name.

Meade are best known for their Schmidt-Cassegrain telescopes (SCTs) and small Maksutov telescopes in the computer-controlled ETX range. These telescope designs are very compact for their size. Most of their instruments are on mountings with the GO TO facility, which – once set up at the beginning of the night's observing – will then slew on request to any object in their database. Synta, however, have concentrated on the more conventional refractors and reflectors, though they also have a range of Maksutovs which will fit on standard mountings.

▼ The Sky-Watcher 100 mm ED refractor on the EQ6 PRO mounting promises to be a popular combination for those who want refractor performance on a mounting capable of taking much larger instruments.

Synta are well known for bringing down the cost of telescopes – and their Sky-Watcher Explorer 130M, mentioned in *Stargazing 2005*, is a popular entry-level reflecting telescope. They have also brought out a range of budget high-quality refractors. Traditionally, refracting telescopes have suffered from the problem of false colour (chromatic aberration) around the edges of bright objects. This is a consequence of the way a simple lens acts as a prism, dispersing light into its component colours. Overcoming it requires the use of an expensive type of glass that refracts the colours in such a way as to counteract the normal dispersion into a rainbow. Telescopes using such lenses are known variously as fluorite refractors, apochromats (APOs) or ED (for extra low dispersion) refractors.

Fluorite refractors are highly prized among astronomers for their exquisite images and wide fields of view. A refracting telescope gives notably better contrast than a reflector, and is also less bother as the optics require much less maintenance and realignment than those of a reflector. But even a modest-sized apochromatic refractor usually costs as much as a large widescreen plasma TV. Then along came Synta with an 80 mm ED refractor costing a quarter of the price for the tube only (known as OTA, for Optical Tube Assembly). Reviews showed that it had excellent optical quality, virtually indistinguishable from much more expensive instruments. Synta have now introduced Sky-Watcher ED 100 mm and 120 mm refractors, with the 100 mm costing less in real terms than the equivalent traditional refractor would have cost a quarter of a century ago. These, and a new 150 mm Maksutov, are in their new PRO-Series range of instruments, which they claim have additional quality control standards.

GO TO wars

Synta have also introduced their first GO TO mountings as upgrades of their heavy-duty Sky-Watcher EQ6 and HEQ5 mounts. As many amateur astronomers have discovered, a good mounting is as important as the optical quality of the telescope, so these could be popular among the more advanced observers. But it remains to be seen whether Synta will take on Meade in the GO TO stakes by bringing out GO TO mounts at entry level.

Meade's GO TO mounts use a system whereby to begin observing you simply point the telescope tube level and north. All that the user then needs to do is enter the date, time and location into the handset and the telescope will automatically locate bright stars, which the user must centre in the field of view. This aligns the telescope correctly for that viewing location. In theory, then, the telescope does most of the work for you, and even a novice should be able to set it up.

Recent models even have electronic systems for levelling and finding north, as well as GPS for fixing the location and time in the more expensive versions. Meade patented the basic 'level and north' idea, which means that other GO TO manufacturers have to either pay Meade a royalty for the use of the principle or devise another method, which invariably demands a greater knowledge of the sky. The Sky-Watcher GO TO, for example, requires the user to know and locate the initial bright stars. So Synta may have a problem in making their systems as potentially user-friendly as those of Meade, though many beginners do find even the Meade GO TO method a challenge to use. Meade announced a loss during 2005, so Synta may arguably now be the leading telescope manufacturer.

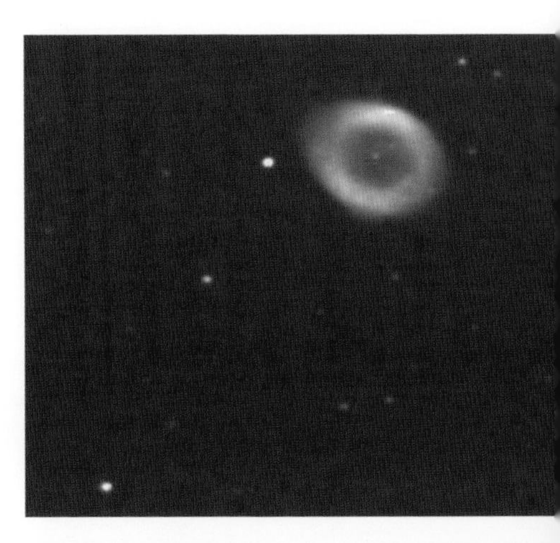

▼ Planetary nebula M57 – the Ring Nebula in Lyra – photographed using a 200 mm telescope with the Meade Deep Sky Imager. The total exposure time was 45 seconds. Stars as faint as 15th magnitude are shown.

Imaging made easy

Meade have put a great deal of effort into making astronomy simpler. They now include an entry-level electronic imaging device, the Deep Sky Imager, with their more expensive telescopes. At around the cost of a reasonable digital camera, the DSI will provide attractive colour images of the brighter deep sky objects with virtually any

▶ *The Deep Sky Imager requires a computer adjacent to the telescope. The image displayed is of the globular cluster M13.*

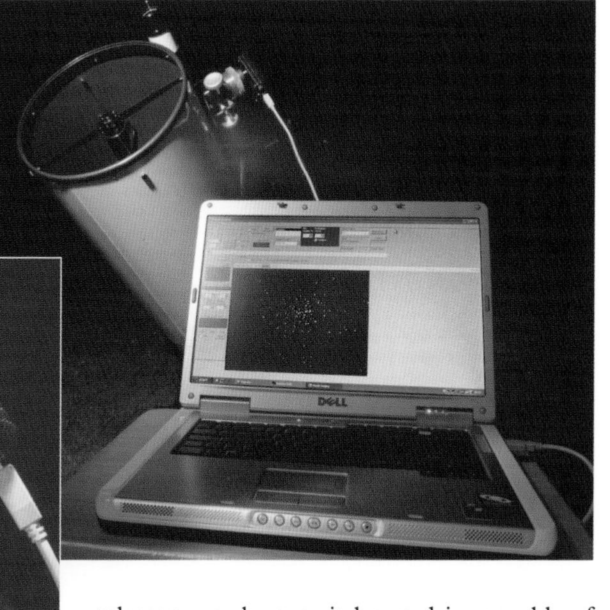

▲ *Meade's Deep Sky Imager – the black box at right – fits in place of the telescope eyepiece and gives live colour images that build up over subsequent exposures to give improved results.*

telescope, as long as it has a drive capable of keeping an object reasonably well centred in the field of view. It does this by adding together fairly short exposures – of the order of 15 seconds – and keeping track of a chosen star within the field of view. Even if the telescope drive is not perfect, the software will detect the shift in the star's position and will realign the subsequent exposures before adding them together. In this way, combined exposure times of several minutes are possible. Normally, this would require careful monitoring of the object during the exposure and frequent correction of the drive rate.

The DSI has its limitations: the imaging chip is not cooled, so there is a limit on the maximum exposure time set by the electronic noise inherent in the chip. But many people are happy just to get their own images of the objects they see through the telescope, and the relatively long exposure times will bring out details and colours in deep sky objects that are invisible to the eye.

The unit does require the user to have a computer with a USB port at the telescope. It is also capable of acting as an autoguider for the telescope for longer exposures. However, this currently requires the computer to have a serial port, which rules out the majority of modern laptops

Big Dobsonians

There is no substitute for a large telescope. Although many people prefer compact instruments that can be carried or stored easily, when it comes to seeing those faint galaxies and nebulae – deep sky objects – aperture wins every time. Objects that can only be seen on the darkest country nights with smaller telescopes

WEBSITES FOR UK SUPPLIERS

Sky-Watcher (Optical Vision): www.opticalvision.co.uk

Meade and Revelation (Telescope House):

www.telescopehouse.co.uk

Celestron and Baader (David Hinds):
www.dhinds.co.uk

become visible from the suburbs when an aperture of 200 to 300 mm is used.

The Dobsonian type of telescope has long been the most affordable means of getting the most for your money. Dobs are basic reflecting telescopes on simple, no-frills mountings. The original intention of the design was to enable an observer to easily build his or her own large telescope using basic materials, having either made or bought the mirrors. But many manufacturers have produced Dobs, and they enjoy great popularity among devotees of deep sky objects.

A range of budget Dobsonians has now been introduced by UK importer Telescope House, also known as Broadhurst, Clarkson & Fuller. These instruments are Chinese-made, though not by Synta, and they have brought the price of a ready-built Dobsonian down to that of the usual price for the mirrors alone. They are available in 200 mm, 250 mm and 300 mm apertures, and even the largest costs about the same as a large domestic TV.

▲ The Revelation 12 Dobsonian from Telescope House provides excellent views of deep sky objects at a budget price. The tube weighs 20 kg and can be lifted onto the mounting quite easily.

Tests of this model show that the performance is excellent for its purpose. A Dobsonian is not intended for detailed planetary views, and indeed the basic mounting is not appropriate for tracking a planet or indeed any other object at high magnification, so there is little point in demanding the highest quality of mirror. However, the mirror quality is good enough to show a considerable amount of detail on the Moon and Jupiter – far more than one would see using a small compact telescope with GO TO, costing the same – while at the same time giving clear views of elusive objects such as the Owl Nebula or, using a filter, the Veil Nebula in Cygnus.

The telescope comes with a good finder, a 32 mm eyepiece with the large 2-inch fitting, and a Crayford-style eyepiece mount that allows smooth focusing. At this price level, many amateurs could afford this instrument as a second telescope to satisfy their thirst for faint fuzzies.

► A webcam view of the Laplace Promontory on the Moon taken through the Revelation 12. Though the telescope is undriven, the video output from the webcam provides several hundred images in a few seconds from which separate photographs can be assembled.